Current Topics in Microbiology

171 and Immunology

Editors

R. W. Compans, Birmingham/Alabama · M. Cooper,
Birmingham/Alabama · H. Koprowski, Philadelphia
I. McConnell, Edinburgh · F. Melchers, Basel
V. Nussenzweig, New York · M. Oldstone,
La Jolla/California · S. Olsnes, Oslo · M. Potter,
Bethesda/Maryland · H. Saedler, Cologne · P. K. Vogt,
Los Angeles · H. Wagner, Munich · I. Wilson,
La Jolla/California

Retroviral Insertion and Oncogene Activation

Edited by H. J. Kung and P. K. Vogt

With 18 Figures

Springer-Verlag

Berlin Heidelberg NewYork
London Paris Tokyo
Hong Kong Barcelona
Budapest

Hsing-Jien Kung

Dept. of Molecular Biology and Microbiology
Case Western Reserve University
School of Medicine
2019 Adelbert Rd.
Cleveland, OH 44106, USA

Peter K. Vogt

Dept. of Microbiology
University of Southern California
School of Medicine
2011 Zonal Avenue HMR-401
Los Angeles, CA 90033-1054, USA

Cover Illustration: Insertional activation of c-myc in lymphomas induced by avian leukosis virus (ALV) and reticuloendotheliosis virus (REV). See page 11, Figure 2, for references.

ISBN-13: 978-3-642-76526-1   e-ISBN-13: 978-3-642-76524-7
DOI: 10.1007/978-3-642-76524-7

© Springer-Verlag Berlin Heidelberg 1991
Library of Congress Catalog Card Number 15-12910
Softcover reprint of the hardcover 1st edition 1991

Typesetting: Thomson Press (India) Ltd, New Delhi;

23/3020-543210—Printed on acid-free paper.

Preface

An integrated retrovirus effectively becomes part of the cellular genome, but with the difference that the virus to a large extent retains control over its own expression through nontranslated sequences in the long terminal repeat (LTR). Some retroviruses also code for nonstructural proteins that further regulate proviral expression. Integration changes the cell genome; it adds viral genes, and in the case of transducing retroviruses also adds cell-derived oncogenes that have been incorporated into the viral genome. Integration can also have consequences for cellular genes. The transcriptional signals in a provirus can activate expression of neighboring cellular genes; the integration even can disrupt and thus inactivate cellular genes. These effects of retroviral genomes take place in cis; they are referred to as insertional mutagenesis and are the subject of this volume.

Almost 10 years have passed since W. Hayward, S. Astrin, and their colleagues found that in B cell lymphomas of chickens, induced by avian leukosis virus, transcription of the cellular proto-oncogene *myc* was upregulated through the integration of a complete or partial provirus in its vicinity. This landmark discovery suggested a mechanism by which retroviruses that do not carry cellular oncogenes in their genome ("nonacute retroviruses") can cause cancer. It contributed the first evidence for the carcinogen potential of oncogenes that are not part of a viral genome. In the intervening years cis-activation of cellular oncogenes has been found in numerous tumor systems that have a non-acute retrovirus as initiating etiological agent. Insertional activation of cellular oncogenes is now recognized as one of the major mechanisms of retroviral oncogenesis. Among the oncogenes targeted by insertional activation are several that can also function as tumor inducing determinants in transducing, rapidly transforming retroviruses. But the list also includes cellular loci that have never been found incorporated into a retrovirus. The oncogenic potential of these cellular loci was at first only suggested by their regular association with the integrated provirus and their consequent, increased expression in tumor tissues. In the meantime, independent supportive evidence strongly favors a tumor inducing role of these loci, adding further weight to the hypothesis that cis-activation is an important step in oncogenesis.

Almost contemporary with the work of the Hayward et al. came the discovery by N. JENKINS and N. COPELAND, that the dilute coat color mutation of mice was caused by proviral insertion. This and other work set the stage for using retroviruses as insertional mutagens to identify and to map genes of importance in animal development. This volume offers summaries and overviews of these various oncogenic and developmental aspects of insertional mutagenesis.

The book is divided into six chapters. The first chapter presents a general review of retroviral insertional activation of cellular oncogenes, with emphasis on the activation mechanisms. The second chapter reviews the application of retroviral mutagenesis in studying genes involved in mouse embryonal development. The remaining chapters focus on oncogenesis and oncogene activations by retroviruses in specific systems. They include mouse mammary carcinomas (Chap. 3), feline leukemia (Chap. 4), and rodent and human lymphomas (Chap. 5). These chapters provide comprehensive information on activated cellular oncogenes and their normal and aberrant functions.

The authors who have contributed to this volume are leaders in the field who have made seminal contributions. They bring to this volume examples from the forefront of current research as well as an authoritative perspective. We thank them for their hard work and their excellent reviews. Special thanks to Drs. NUSSE, LOCK, and NEIL for the timely submission of their manuscripts and their willingness to update their chapters. We also thank our colleagues who provided reprints, preprints, and helpful discussion

HSING-JIEN KUNG
PETER VOGT

Contents

H.-J. KUNG, C. BOERKOEL, and T. H. CARTER:
Retroviral Mutagenesis of Cellular Oncogenes:
A Review with Insights into the Mechanisms
of Insertional Activation. 1

L. F. LOCK, N. A. JENKINS, and N. G. COPELAND:
Mutagenesis of the Mouse Germline Using Retroviruses   27

R. NUSSE: Insertional Mutagenesis
in Mouse Mammary Tumorigenesis. 43

J. C. NEIL, R. FULTON, M. RIGBY, and M. STEWART:
Feline Leukaemia Virus: Generation of Pathogenic
and Oncogenic Variants 67

P. N. TSICHLIS and P. A. LAZO: Virus-Host Interactions
and the Pathogenesis of Murine
and Human Oncogenic Retroviruses 95

Subject Index 173

List of Contributors

(Their addresses can be found at the beginning of their respective chapters.)

BOERKOEL, C. 1	LOCK, L. F. 27
CARTER, T. H. 1	NEIL, J. C. 67
COPELAND, N. G. . . . 27	NUSSE, R. 43
FULTON, R. 67	RIGBY, M. 67
JENKINS, N.A. 27	STEWART, M. 67
KUNG, H.-J. 1	TSICHLIS, P. N. . . . 95
LAZO, P. A. 95	

Retroviral Mutagenesis of Cellular Oncogenes: A Review with Insights into the Mechanisms of Insertional Activation*

H.-J. KUNG[1], C. BOERKOEL[1], and T. H. CARTER[2]

1	Introduction: Retroviral Integration	1
2	Insertional Inactivation of Cellular Genes	2
3	Insertional Activation of Proto-Oncogenes	3
4	Multiple Mechanisms of Insertional Activation	8
4.1	3' LTR Promotion (Promoter Insertion)	8
4.2	5' LTR Promotion (Readthrough Activation)	8
4.3	Enhancer Insertion	10
5	The Choice of Insertional Activation Mechanism	10
5.1	Host Factors Influencing the Mode of Activation	10
5.1.1	What Determines the Choice of 3' LTR Promotion vs Enhancer Insertion Mechanisms?	12
5.1.2	Why are Proviruses Using the Enhancer Insertion Mechanism Oriented Nonrandomly?	12
5.1.3	What Determines the Choice of 5' LTR Promotion vs 3' LTR Promotion?	13
5.2	Viral Elements Influencing the Mode of Insertional Activation	13
5.3	Unraveling Viral Elements Involved in c-*myc* Activation	14
6	Conclusion	16
	References	18

1 Introduction: Retroviral Integration

Retroviruses may be viewed as naturally occurring, self-propagating mutagens of humans and animals. They exert mutagenic effects by integration into the host genome, a natural step in the viral replication cycle. Significant insight into this process has been provided recently by the successful development of in vitro integration systems (BROWN et al. 1987; BOWERMAN et al. 1989; FUJIWARA and

*Support was provided by USPHS grants CA 39207 and CA46613 and a grant from the Ohio Edison Biotechnology Center (to H.J.K.), a USPHS grant to CWRU for trainees (N.B.), and the American Society of Hematology Stratton-Jafee Hematology Scholar Award (to T.H.C.). T.H.C. Dept. of Medicine, University of Oklahoma Health Sciences Center, Oklahoma City, OK 73190, USA
[1] Department of Molecular Biology and Microbiology, Case Western Reserve University, School of Medicine, Cleveland, OH 44106, USA
[2] Department of Molecular Biology and Medicine, Case Western Reserve University, School of Medicine, Cleveland, OH 44106, USA

CRAIGIE 1989; FUJIWARA and MIZUUCHI 1988) and the identification of IN as the viral product involved in integration (SCHWARTZBERG et al. 1984; DONEHOWER and VARMUS 1984; QUINN and GRANDGENETT 1988; TERRY et al. 1988; ROTH et al. 1988). Excellent reviews on retroviral integration are available (e.g., GRANDGENETT and MUMM 1990; VARMUS and BROWN 1989), so only a brief summary is presented here.

Linear viral DNA, which is synthesized in the cytoplasm by reverse transcriptase, appears to be the immediate precursor for integration. The IN protein removes two bases from the 3'-hydroxy- termini of both strands and, presumably in association with other host factors, attaches the 3' recessed ends to the 5' phosphorylated ends of the host DNA. Cleavage of the individual strands of the host DNA is staggered by a distance (four to six nucleotides) determined by the specific viral enzyme. The four- to six-base gaps are then filled in by repair synthesis during integrative recombination. As a consequence, the viral DNA loses two nucleotides from its termini and the host sequence between the staggered cuts is duplicated. These features are characteristic or retroviral insertions.

While the integration process involves very specifically the termini of the linear viral DNA, integration sites in the host DNA appear to be an ensemble of relatively random sequences (DHAR et al. 1980; HUGHES et al. 1981; MAJORS and VARMUS 1981; SHIMOTOHNO et al. 1980). Early analyses were based on sequencing of a limited number of insertion sites and did not give a precise measurement of the extent of randomness in the choice of host sequences. A more complete analysis of the randomness of retroviral insertion was carried out by C.C. SHIH et al. (1988) using a genomic library consisting primarily of LTR (long terminal repeat) cellular junction fragments. These authors found that 75% of the insertion sites were different from one another and thus represent nonspecific insertions. Some 25% of host sites, however, were repeated targets for insertions. Sequences of two such preferred insertion sites had no homology with each other or with the LTR termini. These results reinforced the earlier notion that the choice of retroviral insertion sites is dictated not by primary sequence but by other factors (possibly chromatin structure). Indeed, preferential insertion near transcriptionally active, DNase I- sensitive regions has been reported (KING et al. 1985; ROHDEWOHLD et al. 1987; VIJAYA et al. 1986; MOOSLEHNER et al. 1990).

In summary, retroviruses integrate into the host genome with high efficiency and with little selectivity for host sequences. These properties, together with their high efficiency of replication and infection, make retroviruses powerful insertional mutagens for eukaryotes.

2 Insertional Inactivation of Cellular Genes

Retroviral integration disrupts the host genome and may lead to inactivation of host genes. Retroviral mutagenesis in vivo was demonstrated early in studies of the dilute coat color mutation of mice (JENKINS et al. 1981 and see the chapter by

L.F. LOCK et al., this volume). The phenotype correlates well with the insertion of an endogenous provirus, *Emv-3*, in the chromosome. Reversion to wild-type coat color is associated with loss of the entire viral genome except a single LTR, presumably through homologous recombination between the flanking LTRs (COPELAND et al. 1983; SEPARACK et al. 1988). Other examples include the hairless mutation of mice (STOYE et al. 1988; FRANKEL et al. 1989b) and the slow feathering mutation of chickens (BACON et al. 1989); the genes and inactivation mechanism have yet to be identified, although there is a strong association between the mutant phenotype and the presence of an additional provirus. A homozygous lethal phenotype of mice is caused by Mo-MLV (Moloney murine leukemia virus) insertions near the collagen type I gene (SCHNIEKE et al. 1983), in which insertion in the first intron interferes with the splicing of the gene in a tissue-specific manner (KRATOCHWIL et al. 1989). Retrovirus infection of embryonic stem cells, followed by implantation into pseudopregnant mice to generate mutant transgenics, was proposed as a general "transposon-tagging" method to identify genes responsible for developmental lesions. Spontaneous germ line infection also can induce high rates of insertional mutagenesis in certain strains of mice (SPENCE et al. 1989).

The usefulness of retroviruses as in vitro insertional mutagens for known target genes has been shown in several instances. In an elegant study, VARMUS et al. (1981) introduced a single copy of v-*src* into a cell line, providing a target gene that conferred a transformed phenotype (i.e., anchorage-independent growth). Upon superinfection by Mo-MLV, revertants were selected by bromo-deoxyuridine or hydroxyurea treatment, which kill transformed cells. Reversion caused by Mo-MLV provirus insertion near the *src* gene was shown for two independent mutants. Retroviral mutagenesis has also been demonstrated by using an X-linked gene, *hprt*, as a target (KING et al. 1985). Infection by Mo-MLV and selection for *hprt* mutants resulted in insertional mutation of this gene with a frequency of 10^{-6}–10^{-7}. Similarly, insertion mutations affecting polymorphic surface antigens can be selected in vitro using monoclonal antibodies (FRANKEL et al. 1989). This approach was extended to embryonic stem cells, used to generate a strain of mice to serve as a model for Lesch-Nyhan syndrome (KUEHN et al. 1987). These studies demonstrated the utility of using retroviruses as mutagens. A major limitation of this approach, however, is its unsuitability for autosomal recessive mutations.

3 Insertional Activation of Proto-Oncogenes

Unlike the relative rarity of inactivating mutations, numerous circumstances have been described in which retroviral insertion results in the activation of host genes (NUSSE 1986; CLURMAN and HAYWARD 1988; STAVENHAGEN and ROBINS 1988). Most examples of insertional activation involve somatic mutations of proto-oncogenes which trigger neoplastic transformation. Dominant oncogenes are readily isolated in a diploid background by selection for tumor growth. Strong transcriptional promoter/enhancer elements, as well as translational initiation signals, and other regulatory elements, provide retroviruses with a potent capability to activate gene expression. The first example was reported by HAYWARD and his colleagues in 1981, in which the c-*myc* proto-oncogene was activated in a chicken bursal lymphoma by the insertion of avian leukosis virus (ALV) DNA. Since this seminal discovery, numerous cases of insertional activation of proto-oncogenes and growth-related genes have been identified in cancer tissues or transformed cell lines. Table 1 lists such common insertion sites. Some of the activated loci encode homologues of viral oncogenes, previously identified as a consequence of gene capture by transducing retroviruses. Other loci represent genes that were unknown before their discovery by retroviral insertional activation; most genes identified by these insertion sites have been shown to be active in transformation assays or to have hallmarks of molecules participating in growth-signaling pathways.

Several features of Table 1 are noteworthy: First, exogenous and endogenous viruses, as well as intracisternal A particles, are capable of activating cellular oncogenes. Second, different viruses can induce similar tumors by interacting with the same oncogene, e.g., Rous-associated virus-1 (RAV-1), chicken syncytial virus (CSV), or ring-necked pheasant virus (RPV) induce B-cell lymphomas by activating c-*myc*. Third, a single virus can induce different kinds of neoplasia by interacting with different oncogenes, e.g., RAV-1 can induce B-cell lymphoma, erythroblastosis, nephroblastoma, etc., by activating c-*myc*, c-*erb*-B, c-*fos* or c-Ha-*ras*. These results indicate that tissue-specific transformation is mediated by activation of specific oncogenes. Furthermore, the oncogenic spectrum of a nonacute retrovirus is determined not only by the viral tissue tropism (due to viral envelope or LTR enhancers; reviewed by COFFIN 1990; FAN 1990), but also by the particular oncogenes that the virus is capable of activating. Fourth, multiple proto-oncogenes (or, more accurately, common insertion sites) can be identified in a given type of tumor. For example, in T-cell lymphomas of the mouse, c-*myc*, *pim*-1, and *pim*-2 are potential oncogenes. Likewise, in mouse mammary tumor virus (MMTV) induced mammary carcinomas, *Wnt* 1/*int*-1, *int*-2, and *int*-3 are common insertion sites (see R. NUSSE, this volume). These findings suggest that any one of several oncogenes may trigger the development of T-cell lymphoma or mammary carcinoma. Multiple proto-oncogenes occasionally are activated in the same tumor cell, suggesting that they are cooperating oncogenes or that they act in sequential steps of tumor progression. In other cases, the involvement of different oncogenes in different tumor isolates may be a reflection of genes acting on target cells at specific stages of differentiation in a given lineage. In summary, the identification of new proto-oncogenes by retroviral tagging provides insight into oncogenesis and the normal pathways of cellular growth and differentiation.

Table 1. Common insertion sites in retrovirally mediated oncogenesis

Proto-oncogene (class)[a]	Retroviruses[b]	Neoplasia(species)[c]	Preferred proviral configuration[d]	Predominant activating mechanism[e]	Reference[f]
c-myc (N)	RAV-1/RAV-2	B lymphoma (ch)	S(5')	3'P	1
	CSV	B lymphoma (ch)	S(5')	3'P	2
	RPV	B lymphoma (ch)	S(5')	3'P	3
	CSV	T lymphoma (ch)	S(5'), AS(5')	3'P, E	4
	Mo-MLV/MCF-MLV	T lymphoma (m)	AS(5'), S(3')	E	5
	Mo-MLV	T lymphoma (r)	AS(5'), S(3')	E	6
	FeLV	T lymphoma (c)	AS(5')	E	7
	F-MLV	Erythroleukemia (m)			8
N-myc (N)	Mo-MLV	T lymphoma (m)	AS(5')	E	9
	MLRV	Macrophage hybridomas	S(C)		10
c-erbB (RK)	RAV-1	Erythroblastosis (ch)	S(C)	5'P	11
c-myb (N)	RAV-1/EU8	B lymphoma (ch)	S(C)	5'P	12
	Mo-MLV	Myeloid leukemia (m)	S(C)	5'P	13
c-Ha-ras (GP)	MAV	Nephroblastoma (ch)	S(5')	3'P or 5'P	14
c-Ki-ras (GP)	Mo-MLV	Myelomonocytic leukemia line DA-2 (m)	S(5')	3'P	15
	F-MLV	Mycloid line 416B	S(5')	3'P	15
c-mos (PK)	IAP	Plasmacytoma (m)	AS(5')		16
c-fos (N)	RAV-1	Nephroblastoma (ch)			17
c-rel (N)	ALV	Lymphoma (ch)	S(5')	3'P	17
c-fms/Fim-2 (RK)	F-MLV	Myeloblastic leukemia (m)	AS(5')	E	18
c-lck (PK)	Mo-MLV	T-cell line	S(C)	5'P	19
IL-2 (GF)	GaLV	T-leukemic cell (a) line MLA144	S(3')	E	20
IL-3 (GF)	IAP	Myelomonocytic leukemia (m)			21
GM-CSF (GF)	IAP	Myeloid precursor cell line DIND1 (m)	S(3')	E	22
	R-MLV	Myeloid precursor cell line DIND4 (m)	AS(5')	E	22
	SFFV	Myeloid precursor cell lines DIND5,9 (m)	AS(5')	E	22
c-bic	RAV-1/UR2AV	B lymphoma (ch)	S or AS		23
c-brav-o	RAV-O	B lymphoma (ch)	S or AS		24
Mlvi-1/Mis-1/pvt-1	Mo-MLV	T lymphoma (r)	S or AS (5')	E	25
	Mo-MLV	B lymphoma (m)			26

Table 1 (Continued)

Table 1. (Continued)

Proto-oncogene (class)[a]	Retroviruses[b]	Neoplasia(species)[c]	Preferred proviral configuration[d]	Predominant activating mechanism[e]	Reference[f]
Mlvi-2	Mo-MLV	T lymphoma (r)			27
Mlvi-3	Mo-MLV	T lymphoma (r)			27
Mlvi-4	Mo-MLV	T lymphoma (r)	S (5' or C)	5'P	28
Tpl-1 (N)	Mo-MLV	T lymphoma (r)			29
Gin-1	G-MLV	T lymphoma (m)	S or AS		30
Pim-1 (PK)	Mo-MLV	T lymphoma (m)	S(3')	E	31
	Mo-MLV*	B lymphoma (m)			32
	F-MLV	Erythroleukemia (m)			8
pim-2	Mo-MLV	T lymphoma (m)			33
Dsi-1	Mo-MLV	T lymphoma (r)			34
Ahi-1	Mo-MLV	Pre-B lymphoma (m)	S or AS		35
Fis-1	F-MLV	T lymphoma Myeloid leukemia (m)			36
Fim-1	F-MLV	Myelobastic leukemia (m)	S or AS		18
Fim-3/CB-1	F-MLV	Myeloblastic leukemia (m)	S or As		37
	Mo-MLV	Myeloid leukemia (m)			38
Evi-1 (N)	AKXD-MLV	Myeloid tumors, (m)	S(5', C)	5'P,3', E	39
Evi-2	BXH2-MLV	Myeloid tumors (m)			40
spi-1/sfpi-1 (N)	SFFV/R-MuLV	Erythroleukemia (m)	S(3')	E	41
fli-1 (N)	SFFV	Erythroleukemia (m)	S or AS		42
fhvi-1	FeLV	Spleen lymphoma (c)	S or AS		43
int-1/Wnt-1 (GF)	MMTV	Mammary carcinoma (m)	AS(5'), S(3')	E	44
int-2 (GF)	MMTV	Mammary carcinoma (m)	AS(5'), S(3')	E	45
int-3 (RK)	MMTV	Mammary carcinoma (m)	S(C)		46
int-4/Wnt-3 (GF)	MMTV	Mammary carcinoma (m)	S and AS (5')	E	47
hst/k-FGF (GF)	MMTV	Mammary carcinoma (m)	S(5',3')		48
int-41	MMTV	Mammary carcinoma (m) Kidney adenoma			49
int-H	MMTV	Hyperplastic alveolar nodule (m)			50
p53 (N)	F-MLV	Erythroleukemia (m)			51

a Class of oncogene (if known). GF, growth factor(-like); GP, GTP-binding protein; N, nuclear protein having oncogenic activity or features of a DNA-binding protein; PK, protein kinase (nonrecepto'); RK, receptor(-like) kinase.
b Retroviruses: RAV-1, 2, 0, Rous associated virus 1, 2, 0; CSV, chicken syncytial virus; RPV, ring-necked pheasant virus: Mo-MLV, Moloney murine leukemia virus; Mo-MLV*, Mo-MLV with LTR altered by replacing endogenous enhancer with SV40 enhancer; MLRV, endogenous Moloney-like retrovirus; MCF-MLV, mink

cell focus-forming murine leukemia virus; FeLV, feline leukemia virus; MAV, myeloblastic associated virus; IAP, intracisternal A particle; F-MLV Friend murine leukemia virus; ALV, avian leukosis virus; UR2AV UR2-associated virus; GaLV, Gibbon ape leukemia virus; R-MuLV, Rauscher murine leukemia virus; SFFV, spleen focus-forming virus; G-MLV, Gross murine leukemia virus; MMTV, mouse mammary tumor virus.

c Species: ch, chicken; m, mouse; r, rat; c, cat; a, ape.

d Preferred proviral configuration. S, sense, in the same transcriptional direction as the proto-oncogene; AS, antisense, in the opposite direction to the proto-oncogene; S or AS, proviruses are oriented in one direction (but the relative location to host gene is undetermined); S and AS, proviruses are oriented in both directions; 5', located in the 5' noncoding region of the proto-oncogene; C, located within the coding region of the proto-oncogene; 3', located in the 3' noncoding region of the proto oncogene

e Predominant activating mechanism. 3'P, 3' LTR promotion; 5'P, 5' LTR promotion; E, enhancer insertion.

f *References:*

1. Hayward et al. 1981; Neel et al. 1982; Payne et al. 1982; Fung et al. 1981, 1982; Westaway et al. 1984; Linial and Groudine 1985; Robinson and Gagnon 1986. 2. Noori-Daloii et al. 1981. 3. Simon et al. 1984. 4. Isfort et al. 1987. 5. Corcoran et al. 1984; Selten et al. 1984; Cuypers et al. 1984; Li et al. 1984; O'Donnell et al. 1985; Reicin et al. 1986. 6. Steffen 1984; Tsichlis et al. 1990. 7. Levy et al. 1984; Neil et al. 1984; Forrest et al. 1987; Miura et al. 1989. 8. Dreyfus et al. 1990. 9. Van Lohuizen et al. 1989; Boiocchi et al. 1990. 10. Setoguchi et al. 1989. 11. Fung et al. 1983; Nilsen et al. 1985; Miles and Robinson 1985. 12. Kanter et al. 1988; Pizer and Humphries 1989. 13. Mushinski et al. 1983; Shen-Ong et al. 1984, 1986; Rosson et al. 1987; Gonda et al. 1987. 14. Westaway et al. 1986; see also Pecenka et al. 1988; Soret et al. 1989. 15. Ihle et al. 1989; see also Pecenka et al. 1988; Trusko et al. 1989. 16. Cohen et al. 1983; Gattoni-Celli et al. 1983. 17. Collart et al. 1990; Kabrun et al. 1990. 18. Sola et al. 1988; Gisselbrecht et al. 1987. 19. Voronova and Sefton 1986; Adler et al. 1988. 20. Chen et al. 1985. 21. Dührsen et al. 1990. 22. Stocking et al. 1988. 23. Clurman and Hayward 1989. 24. D. Robinson, L. Crittenden, E. Smith, S. Hughes and H.J. Kung, manuscript in preparation. 25. Tsichlis et al. 1985. 26. Matthews et al. 1989. 27. Tsichlis et al. 1985. 28. Tsichlis et al. 1989, 1990; Lazo et al. 1990. 29. Bear et al. 1989. 30. Villemur et al. 1987. 31. Cuypers et al. 1984; Selten et al. 1985. 32. Hanecak et al. 1988. 33. Breuer et al. 1989. 34. Vijaya et al. 1987. 35. Poirier et al. 1988. 36. Silver and Kozak 1986. 37. Bordereaux et al. 1987. 38. Bartholomew et al. 1989. 39. Mucenski et al. 1988a, b; Morishita et al. 1988. 40. Buchberg et al. 1988, 1990a; 41. Moreau-Gachelin et al. 1988, 1989, 1990; Paul et al. 1989. 42. Ben-David et al. 1990. 43. Levesque et al. 1990. 44. Nusse and Varmus 1982, Nusse et al. 1985, 1990. 45. Peters et al. 1983, 1989a; Dickson et al. 1988, 1990. 46. Gallahan and Callahan 1987. 47. Roelink et al. 1990; R. Nusse, this volume. 48. Peters et al. 1989b. 49. Garcia et al. 1986. 50. Gray et al. 1986. 51. Ben-David et al. 1988; Hicks and Mowat 1988; Dreyfus et al. 1990

4 Multiple Mechanisms of Insertional Activation

Insertional activation of proto-oncogenes generally involves altered trans-cription, although altered processing or stability of transcripts may occasionally have a role. Three major mechanisms for enhanced or deregulated transcription have been identified. These are described below.

4.1 3' LTR Promotion (Promoter Insertion)

Tumors manifesting 3' LTR promotion (Fig. 1a) characteristically have the provirus inserted at sites clustered at the 5' end of the proto-oncogene. The proviruses are oriented in the same transcriptional direction as the gene. The transcript for the activated oncogene begins with the R and U5 regions of the 3' LTR, but contains no other viral sequences. The involved proviruses frequently suffer large deletions of the viral sequences (NEEL et al. 1981; PAYNE et al. 1981; FUNG et al. 1981; SWIFT et al. 1987; GOODENOW and HAYWARD 1987; ROBINSON and GAGNON 1986).

A curious but consistent finding is that the 5' LTR of these proviruses direct transcription at very low or nondetectable levels (SWIFT et al. 1987; GOODENOW and HAYWARD 1987). This pattern of transcriptional activity contrasts significantly with that of intact proviruses, in which the 5' LTR directs strong transcription of the viral genes and the 3' LTR serves primarily as a polyadenylation site. In intact proviruses, the promoter strength of the 3' LTR is estimated to be less than 2% of the 5' LTR (HERMAN and COFFIN 1986). The elements involved in 5' LTR shutoff and 3' LTR turn-on by viral deletions will be discussed in a later section.

4.2 5' LTR Promotion (Readthrough Activation)

This mechanism shares some characteristics with 3' LTR promotion (e.g., the clustering of the proviruses at the 5' end of the gene), but the proviruses involved in activation are often intact and the oncogene transcripts usually contain viral leader sequences (for examples see RAINES et al. 1985; NILSEN et al. 1985; and Table 1). In contrast to 3' LTR promotion these proviruses follow the natural tendency to use the 5' LTR to initiate transcription. Transcription extends through the viral genome as well as through cellular sequences downstream from the insertion site. Most of the primary transcripts are cleaved near the polyA signal of the 3' LTR to generate viral genomic RNA and spliced viral messenger RNAs. Approximately 10%–15% of primary transcripts, however, escape polyadenylation at the 3' LTR site and use a polyA signal of the downstream proto-oncogene (NILSEN et al. 1985). Consequently, a readthrough transcript is generated encompassing both the viral genes and the proto-oncogene (Fig. 7b, c). Splicing a donor site (native or cryptic) of the virus to an acceptor site

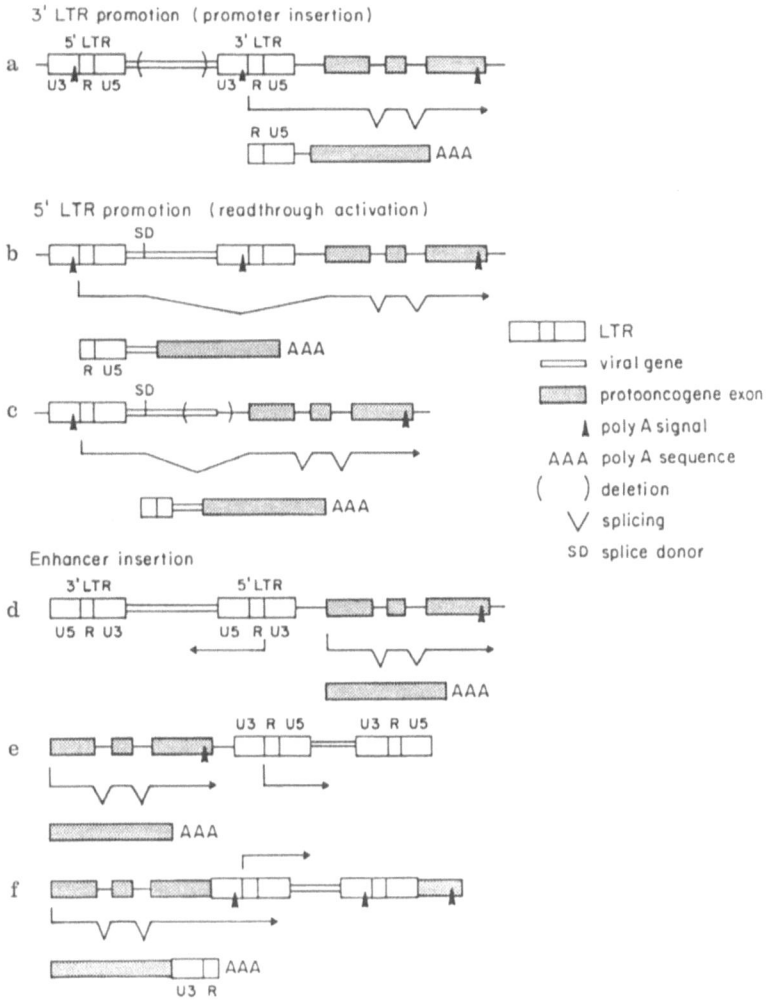

Fig. 1 a–f. The mechanisms of insertional activation of proto-oncogenes. Arrow indicates the direction of transcription; other symbols are defined in the diagram. The diagrams with AAA at the end represent the mature transcripts of the activated proto-oncogenes

of the adjacent proto-oncogene can result in a functional mRNA for an oncogenic fusion protein.

DNA for viral 3′ sequences including the 3′ LTR is sometimes deleted (Fig.1c; ADLER et al. 1988; LINIAL and GROUDINE 1985). Such deletions presumably increase the level of the proto-oncogene message by eliminating interference due to polyadenylation at the 3′ LTR site.

4.3 Enhancer Insertion

Retroviral LTRs carry a number of enhancer motifs, known to bind cellular transactivators. Insertion may introduce enhancer effects on a nearby endogenous promoter. The proviruses involved in enhancer insertion do not have as stringent a position and orientation requirement as in LTR promotion. Thus, insertions may be located upstream or downstream of an activated gene; the provirus may be in the opposite transcriptional orientation and the insertion site may be quite distant from the activated gene. Indeed, insertion at sites up to 300 kb away from a proto-oncogene has recently been reported (TSICHLIS et al. 1990). The activated oncogene transcripts are initiated from natural or cryptic cellular promoters, and the transcript sizes usually are identical to the normal counterpart (see Table 1 for references) (Fig. 1d–f).

When the proviral insertion site is located within the 3' untranslated region of the proto-oncogene and in the same orientation, the transcript is cleaved at the polyA site of the 5' LTR (Fig. 1f). Such truncation of the 3' untranslated sequences of the proto-oncogene might enhance expression at the post transcriptional level. For instance, removal of the AU-rich "RNA instability" element, commonly located near the 3' terminus of short-lived messages (SHAW and KAMEN 1986), may result in elevated steady-state levels of proto-oncogene transcripts.

5 The Choice of Insertional Activation Mechanism

5.1 Host Factors Influencing the Mode of Activation

The structural requirements for activation of a given proto-oncogene clearly dictate the suitability of any mode of proviral activation. With proto-oncogenes, for which truncation of coding sequences contributes to oncogenic conversion, activation generally occurs by 5' LTR promotion. Proto-oncogenes that require simple overexpression for oncogenic conversion generally are activated by 3' LTR promotion and enhancer insertion mechanisms.

Close inspection of different tumor systems reveals that when 3' LTR promotion or enhancer mechanisms are used, the choice between these

───▶

Fig. 2. The insertional maps of the proviruses involved in c-*myc* activation. *Triangles* indicate the insertion sites and orientations of the proviruses in individual tumors. Transcription of c-*myc* is from *left* to *right*; the *open* and *hatched boxes* denote, respectively, the noncoding and coding exons of c-*myc*. (The diagram is a compilation of data taken from the following references: *a*: C.K. SHIH et al. 1984; FUNG et al. 1981; SCHUBACH and GROUDINE 1984; LINIAL and GROUDINE 1985; NOTTENBURG and VARMUS 1986; ROBINSON and GAGNON 1986; *b*: SWIFT et al. 1985. *c*: ISFORT et al. 1987; *d*: CORCORAN et al. 1984; SELTEN et al. 1984; CUYPERS et al. 1984; LI et al. 1984; REICIN et al. 1986. *e*: FORREST et al. 1987; MIURA et al. 1989. *f*: STEFFEN 1984; TSICHLIS et al. 1990)

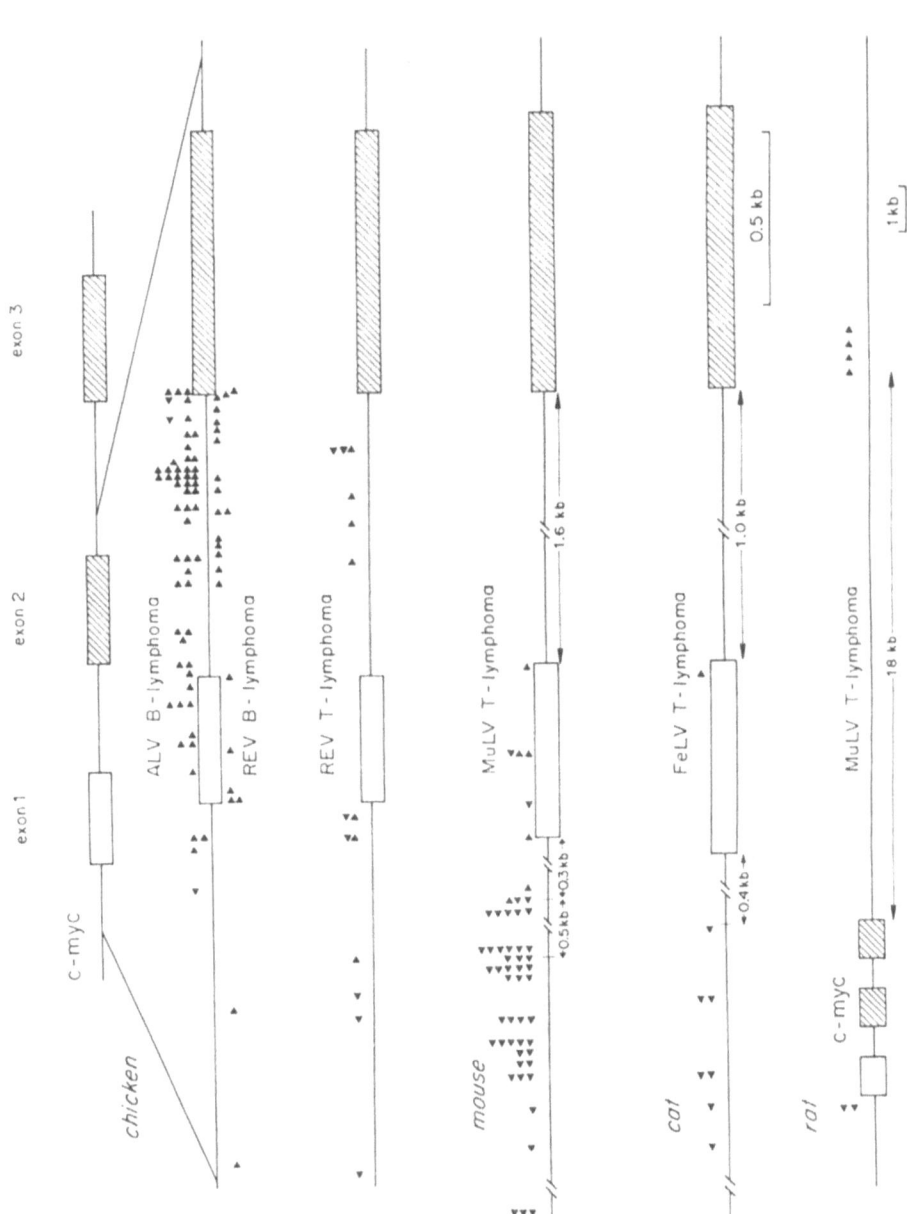

mechanisms is not random. Activation of c-*myc*, which is found in lymphomas of a variety of species, illustrates this point (Fig. 2, Table 1). In avian B-cell lymphomas induced by ALV or CSV strain of reticuloendotheliosis virus (REV), the predominant mode of activation is 3' LTR promotion. Proviral insertion sites are clustered in or before the first intron of c-*myc*; none of these insertions disrupt the protein-coding sequences (which begin in the second exon). In murine or feline T-cell lymphoma, by contrast, enhancer insertion is the exclusive mode for activation of c-*myc*. The proviruses are located either upstream or downstream from c-*myc* and are usually oriented away from the *myc* gene. In avain T-cell lymphomas induced by REV, both 3' LTR promotion and enhancer insertion mechanisms are used. The following discussion offers some spe-culations concerning three major questions raised by these observations.

5.1.1 What Determines the Choice of 3' LTR Promotion vs Enhancer Insertion Mechanisms?

The reasons for the differences in the pattern of insertion used for each model system of c-*myc* activation are not clear. Everything being equal, enhancer insertion (which is relatively independent of position and orientation) should occur most frequently. This is the situation for T-cell lymphomas. Why is 3' LTR promotion strongly favored in avian B-cell lymphoma? Perhaps B-cell lymph-omas require levels of c-*myc* overexpression unattainable by an enhancer mechanism. In B-cell lymphomas, 20- to 50-fold overexpression of the c-*myc* gene is often found (e.g., HAYWARD et al. 1981). This high level of overexpression is likely to be achieved only by 3' LTR promotion. By contrast, a relatively low level of overexpression (6- to 12-fold) is observed in avian T-cell lymphomas (e.g. ISFORT et al. 1987). In mouse T-cell lymphomas, the level of c-*myc* remains almost the same as in the normal T cell; it was suggested that deregulated transcription of the c-*myc* gene by the p1(as opposed to p2) promoter is the critical derangement leading to transformation (REICIN et al. 1986). Additional issues may be involved in the choice of activation mechanisms, such as the predilection of a given virus to undergo the deletions required for 3' LTR promotion.

5.1.2 Why are Proviruses Using the Enhancer Insertion Mechanism Oriented Nonrandomly?

Since enhancers are classically orientation independent, it is curious that the orientation of the involved proviruses is often directed away from the c-*myc* locus. A variety of considerations might select against the less favored orientation. For insertion sites located upstream from c-*myc*, perhaps orientation of the provirus in the same transcriptional direction as c-*myc* would interfere with c-*myc* expression. Such interference might result from promoter occlusion by the strong 5' LTR promoter (see Sect. 5.2) or the interposition of the viral promoter between the viral enhancer and the c-*myc* promoter. For insertion sites located downstream from c-*myc*, orientation of the provirus toward c-*myc* might interfere with *myc* expression by "polymerase collision" between converging transcripts. Additionally, for the special case of insertion sites in the 3' untranslated region of c-*myc* only an orientation away from the bulk of the c-*myc* sequence would provide a polyA signal that might remove an RNA instability element.

5.1.3 What Determines the Choice of 5' LTR Promotion vs 3' LTR Promotion?

Both 5' LTR and 3' LTR promotion can lead to overexpression of proto-oncogene sequences. With proto-oncogenes for which truncation of coding sequences contributes to oncogenic conversion, however, 5' LTR promotion permits the use of the viral leader to generate a functional mRNA for an oncogenic protein. For example, products of certain proto-oncogenes such as growth factor receptors (c-*erb*-B) or nuclear factors (c-*myb*) contain N-terminal inhibitory domains. Activation occurs by truncation of these domains via retroviral insertions within the protein-coding regions (NILSEN et al. 1985; SHEN-ONG et al. 1986; WEINSTEIN et al. 1986). Such insertion removes the original initiation AUG used for translation of the protein. The 5' LTR promotion mechanism, invariably found in these situations, circumvents the translation problem by providing a viral initiation AUG. RNA splicing between viral donor and cellular acceptor sites results in fusion of viral sequences with host sequences. Alternatively, gene fusion may occur at the DNA level by deletion of 3' viral sequences and adjacent host sequences, without involving readthrough and splicing mechanisms.

There are noteworthy exceptions to these generalizations about the choice of activation mechanism. SETOGUCHI et al. (1989) described insertions within the coding sequence for N-*myc* that truncate carboxy-terminal amino acids and also appear to activate transcription by an enhancer mechanism. Especially remarkable is the occurrence of single proviruses that activate multiple genes (PETERS et al. 1989b; BARTHOLOMEW et al. 1989); in one 270-kb cluster of common insertion sites containing c-*myc*, *Mlvi*-1 and *Mlv*-4, insertion into *Mlvi*-1 activates two members of this cluster and insertion into *Mlvi*-4 activates all three genes (TSICHLIS et al. 1989, 1990; LAZO et al. 1990).

5.2 Viral Elements Influencing the Mode of Insertional Activation

Promoter and enhancer activities of viral LTRs depend on the host cell type (reviewed in FAN 1990) as well as internal viral sequences. There are exceptions in which insertional activation is mediated by a solo LTR or in which LTR-directed promotion is mediated in the antisense direction by cryptic promoters (HOROWITZ et al. 1984; CHEN et al. 1985; DICKSON et al. 1990). For most proviruses, the presence of two LTRs with strong promoters creates a complex transcriptional unit. What determines the relative promoter activities of the 5' and 3' LTRs? A solo LTR without other viral sequences generally can act as a strong promoter capable of transcribing heterologous genes. In the context of an intact provirus, as noted earlier, the two identical LTRs show a 50-fold difference in transcriptional activity. Proviruses involved in 3' LTR promotion of cellular oncogenes, however, show a reversed preference for LTR activity. Presumably the choice of LTR promoter is determined by the structure of the provirus.

One mechanism for promoter dominance would be through an enhancer element in the internal viral sequences of an intact provirus that favors transcription from the 5' LTR relative to the 3' LTR. Several studies indicate that viral elements outside the LTR can control LTR activity. ARRIGO et al. (1987) delineated an enhancer sequence located in the *gag* gene of Fujinami sarcoma virus (FSV) and ALV. Sequences in the leader region of REV have been identified that can be footprinted by cell extracts (HIRANO and WONG 1988). For mutine leukemia virus (MuLV), mutation or deletion of the tRNA primer binding site and adjacent sequences activates 5' LTR activity in embryonal carcinoma cells (TAKETO and SHAFFER 1989; BARKLIS et al. 1986; LOH et al. 1990). The findings in B-cell lymphomas lend support to this thesis. In these tumors, 3' LTR activity is required for the transcription of the c-*myc* gene and 5' LTR activity is low or absent. Deletions of proviral sequences are frequently found near the 5' LTR, which would be consistent with the loss of a 5' LTR-specific enhancer.

An additional mechanism for 5' LTR dominance in intact proviruses might involve transcription from the upstream promoter interfering with the activity of the downstream promoter. This situation would be analogous to the "promoter-occlusion" phenomenon first described in prokaryotes (ADHYA and GOTTESMAN 1982). CULLEN et al. (1984) observed such interference between two LTRs derived from ALV; mutations of the 5' LTR which reduce its promoter activity result in a concomitant rise in 3' LTR transcription of up to three- to nine-fold. Perhaps a combination of 5' LTR-specific enhancer and promoter-occlusion mechanisms would account for the dramatic differential promoter activity seen in intact proviruses.

5.3 Unraveling Viral Elements Involved in c-*myc* Activation

We have studied 5' and 3' LTR promoter activity in the context of a partially deleted provirus originally involved c-*myc* activation of a B-cell lymphoma. The 713 provirus was isolated from a CSV-induced tumor(SWIFT et al. 1987). The provirus has suffered an extensive deletion, which begins immediately 3' to the primer binding site and includes 80% of the genome. The two LTRs are linked by 800 nucleotides from the 3' portion of the provirus and an inversion of a part of *env*. Both LTRs are intact, without even a single base change. In the original tumor the c-*myc* gene is highly expressed from the 3' LTR, but no transcription is detactable from the 5' LTR. Transfection of the deleted provirus and the entire c-*myc* locus into fibroblasts reproduces the predominance of 3' LTR-initiated transcription (C. BOERKOEL and H.-J. KUNG, unpublished data), indicating that the differential activity of the LTRs is not specific to the lymphoid environment.

Replacement of c-*myc* with a growth hormone gene gives similar results, ruling out an effect on transcription specific for c-*myc* sequences. Nuclear run-on experiments show that the differential transcription of the LTRs is most likely at the level of transcription initiation or elongation. Restoration of an 87-element, consisting of the CSV leader sequence contiguous with the primer binding site, increase 5′ LTR activity with a concomitant loss of 3′ LTR activity. The element is active in only one orientation and only when located downstream from the 5′ LTR, in contrast to a classical enhancer. This behavior is similar to the RNA enhancer recently described for the TAR region of human immunodeficiency virus (HIV) (SHARP and MARCINIAK 1989; BERKHOUT et al. 1989; SOUTHGATE et al. 1990). An RNA enhancer would be expected to act on the upstream promoter, thus ensuring the biased activation of the 5′ LTR. It is unclear what role proviral structures may have in modulating the use of alternative promoter or splice sites. Our results show that the intact CSV provirus appears to be equipped with a sequence that can feed back to the upstream LTR, perhaps to enhance or stabilize the RNA polymerase complex. In proviruses mediating activation of c-*myc* by 3′ LTR promotion, the deletion of this directional enhancing element would account partially for the low 5′ LTR and high 3′ LTR promoter activities.

What determines the complete shut off of the 5′ LTR in the 713 provirus? Deletion of the 87 enhancer element would not explain the complete switch of transcriptional activity to the 3′ LTR. With such a deletion the 5′ and 3′ LTRs should have equal strength, and any degree of promoter occlusion would tend to favor the 5′ LTR. One possible explanation would be the presence of an additional enhancer element, situated proximal to the 3′ LTR and with less strength than the 5′ LTR-specific enhancer. In the absence of the 5′ LTR enhancer (as in the case of 713 provirus) such a 3′ enhancer might make the 3′ LTR a stronger promoter. Enhancer elements proximal to the 3′ LTR have previously been identified in the REV system, to which CSV belongs (HIRANO and WONG 1988), and in the ALV system (LAIMINS et al. 1984; NORTON and COFFIN 1987). Additional factors may be invoked to explain the full shut off of the 5′ LTR in the 713 provirus. Sufficient transcription from the 3′ LTR might reduce 5′ LTR activity by promoter competition or epigenetic suppression. These terms have been used to describe the phenomenon of mutual interference by two tandem promoters. They imply bidirectional effects (i.e., interference with the 3′ promoter by the 5′ promoter activity or vice versa), in contrast to the promoter-occlusion model (in which only the 5′ promoter interferes with the 3′ promoter). In promoter competition, as originally described for the yeast HIS4 locus, a mutation activating either one of two tandem promoters downregulated the other one, perhaps, by competing for a limiting transcriptional factor such as TFIID (HIRSCHMAN et al. 1988; EISENMANN et al. 1989). Epigenetic suppression was demonstrated for retroviruses containing an internal promoter (EMERMAN and TEMIN 1984, 1986). Selectable genes were placed under transcriptional control of either the 5′ LTR or the internal promoter. Selection for transcription by the internal promoter downmodulated the gene directed by the 5′ LTR and vice versa. The authors found that not all promoters interfered equally well with each other. While the exact mechanism of epigenetic suppression is not clear, perturbation of chromatin structure of one promoter by the activity of the adjacent one was suggested.

6 Conclusion

In the 10 years since discovery of c-*myc* activation by ALV insertion in B-cell lymphoma (HAYWARD et al. 1981), at least 40 common insertion sites have been identified in various models of retrovirally induced tumors. In nearly all cases in which genes residing in the common insertion sites have been identified, they are either known oncogenes or genes related to molecules involved in growth regulation. Several of the newly identified genes were shown to possess transforming potential in vitro or in vivo (e.g. TSUKAMOTO et al. 1988). Thus, retroviral tagging has been one of the most rewarding approaches in identifying oncogenes. The ways in which proviral insertions can perturb target genes have taught us a great deal about the oncogenic conversion of normal genes. Oncogenic conversion of genes for growth factor-like molecules (*Wnt-1/int-1*, *int-2*II-2, etc.) generally requires only deregulation or overexpression of the gene product; no structural changes of the proteins are necessary. In these cases, retroviral insertions are invariably outside the coding domains (DICKSON et al. 1990). By contrast, genes for growth factor receptors (c-*erb*-B, *int*-3, etc.) or certain transcriptional factors (c-*myb*) may be activated by removal of negative-controlling domains. Retroviral insertions occur within the coding region to truncate the proto-oncogene, providing not only the transcriptional promoter but also an initiation AUG (e.g., NILSEN et al. 1985; SHEN-ONG et al. 1986; ADLER et al. 1988) and, in one instance, a signal peptide of the viral *env* product (NILSEN et al. 1985)—a remarkable feat unparalleled by other mutagens.

Another outcome of these studies is an understanding of the mechanism by which oncogenes may be transduced (i.e., how cellular oncogenes may be captured or incorporated by acute transforming retroviruses) (SWANSTROM et al. 1083; HERMAN and COFFIN 1986; RAINES et al. 1988) It has generally been accepted that transduction begins with insertion of a provirus at a site adjacent to the cellular oncogene. 5′ LTR promotion can lead to the expression of viral and proto-oncogene sequences on a single transcript. Such a transcript could be packaged together with a viral genomic RNA in a (diploid) virion, increasing the chance of further recombination during a second round of infection. Thus, well-documented consequences of retroviral infection could provide for the recombinational steps needed to generate an acute transforming retrovirus. An exceptionally high frequency of *erb* B transduction has been observed in ALV-induced erythroblastosis (MILES and ROBINSON 1985; RAINES et al. 1988). Newly generated erb B-transducing viruses were isolated from leukemic birds. Frequently, these viruses have not undergone extensive genetic changes, which accumulate during prolonged in vitro propagation. The polyA sequence of the original readthrough transcript is present in the genome of two independently isolated transducing viruses (RAINES et al. 1988), supporting the notion that one step of oncogene transduction involves an RNA intermediate (see also HUANG et al. 1986). The feline leukemia viruses also have a penchant for oncogene transduction (e.g. STEWART et al. 1986; J. NEIL, this volume). At least seven oncogenes have been captured by Feline leukemid virus (FeLV), and examination of the viral-oncogene junctions has provided further insight into the recombination process.

In contrast to the frequently described activation of oncogenes, the inactivation of an antioncogene, *p53*, has been described in Friend murine leukemia virus (F-MLV) induced erythroleukemia (WOLF and ROTTER 1984; WOLF et al. 1984; HICKS and MOWAT 1988; CHOW et al. 1987; BEN-DAVID et al. 1988). It initially seemed unlikely that gene-inactivation would arise by two insertional events into both alleles, but independent insertional mutations that knock out both alleles of the *p53* gene have been reported for one tumor cell line (HICKS and MOWAT 1988). A variety of *p53* gene rearrangements, including retroviral insertions or host sequence deletions, have been found in about 20% of tumor cell clones. The majority of these rearrangements are associated with the loss of *p53* expression. It remains unclear how proviral insertion or other rearrangements in one allele of many cell lines may ablate expression of *p53* in the presence of an apparently normal allele (BEN-DAVID et al. 1988). Some rearrangements result in the expression of truncated proteins; perhaps mutant *p53* protein, generated by insertion into one allele, might act in a *trans*-dominant negative fashion. It is now thought that some mutant *p53* proteins can abolish the activity of wild-type *p53*, perhaps by forming an inactive oligomer or by competing for a limiting substrate (FINLAY et al. 1989). Retroviral mutagenesis might be useful for identifying additional antioncogenes and, indeed, some of the currently uncharacterized common insertion sites may eventually prove to be antioncogenes.

Finally, what we have learned about retroviral mutagenesis in the activation of oncogenes suggests ways that cellular genes might be activated by related mutagenic insults. Transposon-associated chromosomal rearrangements causing similar perturbations of gene expression or structure might activate proto-oncogenes. Very limited data hints that retroviral infection might induce deletions (directly or indirectly) without necessarily leaving a transposon tag (VARMUS et al. 1981; BEN-DAVID et al. 1988). Perhaps retrovirally induced disease could involve "hit and run" events, in which some genetic aberrations would lack an identifiable signature of viral integration. In addition, other transposable elements might activate oncogenes by insertional mutagenesis. Long interspersed repeated DNA element (LINE) insertions have been found in the c-*myc* locus in canine transmissible veneral tumors (KATZIR et al. 1985,1987) and a human breast carcinoma (MORSE et al. 1988). LINE involvement has also been found in translocation of c-*myc* in a rat immunocytoma (PEAR et al. 1988). Germline mutations that appear to involve insertional activation of cellular genes include a

LINE-induced polymorphism of the *Mlvi-2* locus possibly associated with thymoma development in the Buffalo rat (ECONOMOU-PACHNIS et al. 1985) and an intracisternal A particle (IAP) insertion in the 3′ region of a mouse gene for renin that appears to cause high levels of renin expression (BURT et al. 1984). In addition, evolution of human amylase genes appears to have involved multiple insertional mutations including an endogenous retrovirus (SAMUELSON et al. 1990). Perhaps in the future we may find a larger role for retroviruses and other transposable elements in mutagenic events that activate cellular genes of vertebrates.

Acknowledgements. The authors thank S. Goff, D. Grandgenett, and J. Leis for providing helpful manuscripts and discussion, and D. Angeloff for expert typing.

References

Adhya S, Gottesman M (1982) Promoter occlusion: transcription through a promoter may inhibit activity. Cell 29: 939–944

Adler HT, Reynolds PJ, Kelley CM, Sefton BM (1988) Transcriptional activation of *lck* by retrovirus promoter insertion between two lymphoid- specific promoters. J Virol 62: 4113–4122

Arrigo S, Yun M, Beemon K (1987) *cis*-Acting elements within the *gag* genes of avian retroviruses. Mol Cell Biol 7: 388–397

Bacon LD, Smith E, Crittenden LB, Havenstein GB (1990) Association of the slow feathering (*K*) and an endogenous viral (*ev21*) gene on the Z chromosome of chickens. Poult Sci 67(2): 191–197

Barklis E, Mulligan RC, Jaenisch R (1986) Chromosomal position or virus mutation permits retrovirus expression in embryonal carcinoma cells. Cell 47: 391–399

Bartholomew C, Morishita K, Askew D, Buchberg A, Jenkins NA, Copeland NG, Ihle JN (1989) Retroviral insertions in the CB-1/*Fim-3* common site of integration active expression of the *Evi-1* gene. Oncogene 4: 529–534

Bear SE, Bellacosa A, Lazo PA et al. (1989) Provirus insertion in *Tpl-* 1, an *Ets*-1 related oncogene, is associated with tumor progression in Moloney murine leukemia virus-induced rat thymic lymphomas. Proc Natl Acad Sci USA 86: 7495–7499

Ben-David Y, Giddens EB, Bernstein A (1990) Identification and mapping of a common proviral integration site *Fli*-1 in erythroleukemia cells induced by Friend murine leukemia virus. Proc Natl Acad Sci USA 87: 1332–1336

Ben-David Y, Prideaux VR, Chow V, Benchimol S, Bernstein A (1988) Inactivation of the *p53* oncogene by internal deletion or retroviral integration in erythroleukemic cell lines induced by Friend leukemia virus. Oncogene 3: 179–185

Berkhout B, Silverman RH, Jeang K-T (1989) Tat *trans*-activates the human immunodeficiency virus through a nascent RNA target. Cell 59: 273–282

Boiocchi M, Doketti R, Maestro R, Feriotto G, Rizzo S, D.Re V, Soriego F (1990) A coordinated proto-oncogene expression characterizes MCF 247 murine leukemia virus-induced T-cell lymphomas irrespectively of proviral insertion affecting *myc* loci. Leuk Res 14: 549–558

Bordereaux D, Fichelson S, Sola B, Tambourin PE, Gisselbrecht S (1987) Frequent involvement of the *fim*-3 region in Friend murine leukemia virus-induced mouse myeloblastic leukemias. J Virol 61: 4043–4045

Bowerman B, Brown PO, Bishop JM, Varmus HE (1989) A nucleoprotein complex mediates the integration of retroviral DNA. Genes Dev 3: 469–478

Breuer ML, Cuypers HT, Berns A (1989) Evidence for the involvement of *pim*-2, a new common insertion site, in progression of lymphomas. EMBO J 8: 743–747

Brown PO, Bowerman B, Varmus HE, Bishop JM (1987) Correct integration of retroviral DNA in vitro. Cell 49: 347–356

Buchberg AM, Bedigian HG, Taylor BA et al. (1988) Localization of Evi-2 to chromosome 11: linkage to other proto- oncogene and growth factor loci using interspecific backcross mice. Oncogene Res 2: 149–165

Buchberg AM, Bedigan HG, Jenkins NA, Copeland NG (1990) Evi- 2, a common integration site involved in murine myeloid lekemogenesis. Mol Cell Biol 10: 4658–4666

Burt, DW, Reith AD, Brammar WJ (1984) A retroviral provirus closely associated with the Ren-2 gene of DBA/2 mice. Nucleic Acids Res 12: 8579–8593

Chen SJ, Holbrook, NJ, Mitchell KF, Vallone CA, Greengard JS, Crabtree GR, Lin Y (1985) A viral long terminal repeat in the interleukin-2 gene of a cell line that constitutively produces interleukin-2. Proc Natl Acad Sci USA 82: 7284–7288

Chow V, Ben-David Y, Bernstein A, Benchimol S, Mowat M (1987) Multistage Friend erythroleukemia: independent origin of tumor clones with normal or rearranged p53 cellular oncogenes. J Virol 61: 2777–2781

Clurman BE, Hayward, WS (1988) Insertional activation of proto-oncogenes previously identified as viral oncogenes. In: Klein G. (ed) Cellular oncogene activation. Dekker, New York pp 55–94

Clurman, BE, Hayward, WS (1989) Multiple proto-oncogene activations in avian leukosis virus-induced lymphomas: evidence for stage-specific events. Mol Cell Biol 9: 2657–2664

Coffin JM (1990) Retroviridae and their replication. In: Fields BN et al. (eds) Virology. Raven, New York, pp 1437–1500

Cohen JB, Unger T, Rechavi G, Camaani E, Givol D (1983) Rearrangement of the oncogene c-mos in mouse myeloma NSI and hybridomas. Nature 306: 797–799

Collart KL, Aurigemma R, Smith RE, Kawai S, Robinson HL (1990) Infrequent involvement of c-fos in avian leukosis virus-induced nephroblastoma. J Virol 64: 3541–3544

Copeland NG, Hutchison KW, Jenkins NA (1983) Excision of the DBA ecotropic provirus in dilute coat-color revertants of mice occurs by homologous recombination involving the viral LTRs. Cell 33: 379–387

Corcoran LM, Adams JM, Dunn AR, Cory S (1984) Murine T lymphomas in which the cellular myc oncogene has been activated by retroviral insertion. Cell 37: 113–122

Cullen, BR, Lomedico PT, Ju G (1984) Transcriptional interference in avian retroviruses—implications for the promoter insertion model of leukaemogenesis. Nature 307: 241–245

Cuypers HTM, Selter G, Quint W et al. (1984) Murine leukemia virus-induced T-cell lymphomagenesis: integration of proviruses in a distinct chromosomal region. Cell 37: 141–150

Cuypers HTM, Selten GC, Zijlstra M, de Goede RE, Melief CJ,Berns AJ (1986) Tumor progression in murine leukemia virus-induced T-cell lymphomas: monitoring clonal selections with viral and cellular probes J Virol 60: 230–241

Dhar R, McClements WL, Enquist LW, Vande Woude GF (1980) Nucleotide sequences of integrated Moloney sarcoma provirus long terminal repeats and their host and viral junctions. Proc Natl Acad Sci USA 77: 3937–3941

Dickson C, Smith R, Brookes S, Peters G (1988) Tumorigenesis by mouse mammary tumor virus: proviral activation of a cellular gene in the common integration region int-2. Cell 37: 529–536

Dickson C, Smith R, Brookes S, Peters G (1990) Proviral insertions within the int-2 gene can generate multiple anomalous transcripts but leave the protein-coding domain intact. J Virol 64: 784–793

Donehower LA, Varmus HE (1984) A mutant murine leukemia virus with a single missense codon in pol is defective in a function affecting integration. Proc Natl Acad Sci USA 81: 6461–6465

Dreyfus F, Sola B, Fichelson S, Varlet P, Charon M, Tambourin P, Wendling F, Gisselbrecht S (1990) Rearrangements of the Pim-1, c-myc, and p53 genes in Friend helper virus-induced mouse erythroleukemias. Leukemia 4: 590–594

Dührsen U, Stahl J, Gough NM (1990) In vivo transformation of factor- dependent hemopoietic cells: role of intracisternal A-particle transposition for growth factor gene activation. EMBO J 9: 1087–1096

Economou-Pachnis A, Lohse MA, Furano AV, Tsichlis PN (1985) Insertion of long interspersed repeated elements at the Igh (immunoglobulin heavy chain) and Mlvi-2 (Moloney leukemia virus integration 2) loci of rats. Proc Natl Acad Sci USA 82: 2857–2861

Eisenmann DM, Dollard C, Winston F (1989) SPT15, the gene encoding the yeast TATA binding factor TFIID, is required for normal transcription initiation in vivo. Cell 58: 1183–1191

Emerman M, Temin HM (1984) Genes with promoters in retrovirus vectors can be independently suppressed by an epigenetic mechanism. Cell 39: 459–467

Emerman M, Temin HM (1986) Comparison of promoter suppression in avian and murine retrovirus vectors. Nucleic Acids Res 14: 9381–9396

Fan H (1990) Influences of the long terminal repeats on retrovirus pathogenicity. Semin virol 1: 165–174

Finlay CA, Hinds PW, Levine AJ (1989) The p53 proto-oncogene can act as a suppressor of transformation. Cell 57: 1083–1093

Forrest D, Onions D, Lees G, Neil JC (1987) Altered structure and expression of c-*myc* in feline T-cell tumours. Virology 158: 194–205

Frankel WN Potter TA, Rajan TV (1989a) Effect of proviral insertion on transcription of the murine *B2m*b gene. J Virol 63: 2623–2628

Frankel WN, Stoye JP, Taylor BA, Coffin JM (1989b) Genetic analysis of endogenous xenotropic murine leukemia viruses: association with two common mouse mutations and the viral restriction locus *Fv*-1 J Virol 63: 1763–1774

Fujiwara T, Craigie R (1989) Integration of mini-retroviral DNA :a cell-free reaction for biochemical analysis of retroviral integration. Proc Natl Acad Sci USA 86: 3065–3069

Fujiwara T, Mizuuchi K (1988) Retroviral DNA integration: structure of an integration intermediate. Cell 54: 497–504

Fung Y-KT, Fadly AM, Crittenden LB, Kung HJ (1981) On the mechanism of retrovirus-induced avian lymphoid leukosis: deletion and integration of the proviruses. Proc Natl Acad Sci USA 78: 3418–3422

Fung Y-KT, Crittenden LB, Kung HJ (1982) Orientation and position of avian leukosis virus DNA relative to the cellular oncogene c-*myc* in B-lymphoma tumors of highly susceptible 151_5X7_2 chickens. J Virol 44: 742–746

Fung, Y-KT, Lewis WG, Crittenden LB, Kung HJ (1983) Activation of the cellular oncogene c-*erb*B by LTR insertion: molecular basis for induction of erythroblastosis by avian leukosis virus. Cell 33: 357–368

Gallahan D, Callahan R (1987) Mammary tumorigenesis in feral mice: identification of a new *int* locus in mouse mammary tumor virus (Czech II)-induced mammary tumors. J Virol 61: 66–74

Garcia M, Wellinger R, Vessaz A, Digglemann H (1986) A new site of integration for mouse mammary tumor virus proviral DNA common to Balb/cf(C3H) mammary and kidney adenocarcinomas. EMBO J 5:127–134

Gattoni-Celli S, Hsiao W-LW, Weinstein IB (1983) Rearranged c-*mos* locus in a MOPC21 murine myeloma cell line and its persistence in hybridomas. Nature 306: 795–796

Gisselbercht S, Fichelson S, Sola B et al. (1987) Frequent c-*fms* activation by proviral insertion in mouse myeloblastic leukaemias. Nature 329: 259–261

Gonda TJ, Cory S, Sobieszczuk P, Holtzman D, Adams JM (1987) Generation of altered transcripts by retroviral insertion in the c-*myb* gene in two murine monocytic leukemias. J Virol 61: 2754–2763

Goodenow MM, Hayward WS (1987) 5' Long terminal repeats of *myc*- associated proviruses appear structurally intact but are functionally impaired in tumors induced by avian leukosis viruses. J Virol 61: 2489–2498

Grandgenett DP, Mumm SR (1990) Unraveling retrovirus integration. Cell 60: 3–4

Gray DA, McGrath CM, Jones RF, Morris VL (1986) A common mouse mammary tumor virus integration site in chemically induced precancerous mammary hyperplasias. Virology 148: 360–368

Hanecak R, Pattengale PK, Fan H (1988) Addition or substitution of simian virus 40 enhancer sequences into the Moloney murine leukemia virus (M-MuL V) long terminal repeat yields infectious M-MuLV with altered biological properties. J Virol 62: 2427–2436

Hayward WS, Neel BG Astrin, SM (1981) Activation of a cellular *onc* gene by promoter insertion in ALV-induced lymphoid leukosis. Nature 290: 475–480

Herman SA, Coffin JM (1986) Differential transcription from the long terminal repeats of integrated avain leukosis virus DNA. J Virol 60: 497–505

Hicks GG, Mowat M (1988) Integration of Friend murine leukemia virus into both alleles of the p53 oncogene in an erythroleukemic cell line. J Virol 62: 4752–4755

Hirano A, Wong T (1988) Functional interaction between transcriptional elements in the long terminal repeat of reticuloendotheliosis virus: cooperative DNA binding of promoter-and enhancer-specific factors. Mol Cell Biol 8: 5232–5244

Hirschman JE, Durbin KJ, Winston F (1988) Genetic evidence for promoter competition in *Saccharomyces cerevisiae*. Mol Cell Biol 8: 4608–4615

Horowitz M, Luria S, Rechavi G, Givol D (1984) Mechanism of activation of the mouse c-*mos* oncogene by the LTR of an intracisternal A-particle gene. EMBO J 3: 2937–2941

Huang CC, Hay, N, Bishop, JM (1986) The role of RNA molecules in transduction of the proto-oncogene c-*fps*. Cell 44: 935–940

Hughes, SH, Mutschler A, Bishop JM, Varmus HE (1981) A Rous sarcoma virus provirus is flanked by short direct repeats of a cellular DNA sequence present in only one copy prior to integration. Proc Natl Acad Sci USA 78: 4299–4303

Ihle JN, Smith-White B, Sisson B et al. (1989) Activation of the c-H- ras proto-oncogene by retrovirus insertion and chromosomal rearrangement in Moloney leukemia virus-induced T-cell leukemia. J Virol 63: 2959–2966

Isfort R, Witter RL, Kung HJ (1987) c-myc Activation in an unusual retrovirus-induced avian T-lymphoma resembling Marek's disease: proviral insertion of 5' of exon one enhances the expression of an intron promoter. Oncogene Res 2: 81–94

Jenkins NA, Copeland NG, Taylor BA, Lee BK (1981) Dilute (d) coat colour mutation of DBA/2J mice is assoicated with the site of integration of an ecotropic MuLV genome. Nature 293: 370–374

Kanter MR, Smith RE, Hayward WS (1988) Rapid induction of B-cell lymphomas:insertional activation of c-myb by avian leukosis virus. J Virol 62: 1423–1432

Katzir N, Arman E, Cohen D, Givol D, Rechavi G (1987) Common origin of transmissible venereal tumors (TVT) in dogs. Oncogene 1: 445–448

Katzir N, Rechavi G, Cohen JB, Unger T, Simoni F, Segal S, Cohen D, Givol D (1985) "Retrotransposon" insertion into the cellular oncogene c-myc in canine transmissible veneral tumor. Proc Natl Acad Sci USA 82: 1054–1058

King W, Patel MD, Lobel LI, Goff SP, Nguyen-Huu MC (1985) Insertion mutagenesis of embryonal carcinoma cells by retroviruses. Science 228: 554–558

Koehne CF, Lazo PA, Alves K, Lee JS, Tsichlis PN, O'Donnell PV (1989) The Mlvi-1 locus involved in the induction of rat T-cell lymphomas and the pvt-1/mis 1 locus are identical. J Virol 63: 2366-2369

Kratochwil K, von der Mark K, Kollar EJ et al. (1989) Retrovirus-induced insertional mutation in Mov13 mice affects collagen I expression in a tissue-specific manner. Cell 57: 807–816

Kuehn MR, Bradley A, Robertson EJ, Evans MJ (1987) A potential animal model for Lesch-Nyhan syndrome through introduction of HPRT mutations into mice. Nature 326: 295–298

Laimins LA, Tsichlis P, Khoury G (1984) Multiple enhancer domains in the 3'terminus of the Prague strain of Rous sarcoma virus. Nucleic Acids Res 12: 6427–6442

Lazo PA, Lee JS, Tsichlis PN (1990) Long-distance activation of the myc protooncogene by provirus insertion in Mlvi-1 or Mlvi-4 in a rat T-cell lymphomas. Proc Natl Acad Sci USA 87: 170–173

Levesque KS, Bonham L, Levy LS (1980) flvi-1 A common integration domain of feline leukemia virus in naturally occurring lymphomas of a particular type. J Virol 64: 3455–3462

Levy LS, Gardner MB, Casey JW (1984) Isolation of a feline leukemia provirus containing the oncogene myc from a feline lymphosarcoma. Nature 308: 853–856

Li Y, Holland, CA, Hartley JW, Hopkins N (1984) Viral integration near c- myc in 10–20% of MCF 247-induced AKR lymphomas. Proc Natl Acad Sci USA 81: 6808–6811.

Linial M, Groudine M (1965) Transcription of three c-myc exons is enhanced in chicken bursal lymphoma cell lines. Proc Natl Acad Sci USA 82: 53–57

Linial M, Gunderson N, Groudine M (1985) Enhanced transcription of c-myc in bursal lymphoma cells requires continuous protein synthesis. Science 230: 1126–1132

Loh TP, Sievert, LL, Scott RW (1990) Evidence for a stem cell-specific repressor of Moloney murine leukemia virus expression in embryonal carcinoma cells. Mol Cell Biol 10: 4045–4057

Majors JE, Varmus HE (1981) Learning about the replication of retroviruses from a single cloned provirus of mouse mammary tumor virus. ICN-UCLA Symposium. Mol Cell Biol 18: 241–253

Matthews EA, Vasmel WL, Schoenmakers HJ, Melief CJ (1989) Retrovirally induced murine B-cell tumors rarely show proviral integration in sites common T-cell tumors. Int J Cancer 43: 1120–1125

Miles BD, Robinson HL (1985) High-frequency transduction of c-erb B in avian leukosis virus-induced erythroblastosis J Virol 54: 295–303

Miura T, Shibuya M, Tsujimoto H, Fukasawa M, Hayami M (1989) Molecular cloning of a feline leukemia provirus integrated adjacent to the c-myc gene in a feline T-cell leukemia cell line and the unique structure of its long terminal repeat. Virology 169: 458–461

Mooslehner K, Karls U, Harbers K (1990) Retroviral integration sites in transgenic Mo mice frequently map in the vicinity of transcribed DNA regions. J Virol 64: 3056–3058

Moreau-Gachelin F, Tavitian A, Tambourin P (1988) Spi-1 is a putative oncogene in virally induced murine erythroleakaemias. Nature 331: 277–280

Moreau-Gachelin F, Ray D, Mattei MG, Tambourin P, Tavitian A (1989) The putative oncogene Spi-1 : murine chromosomal localization and transcriptional activation in murine acute eryth-roleukemias. Oncogene 4: 1449–1456

Moreau-Gachelin F, Ray D, de Both NJ, van der Feltz MJ, Tambourin P, Tavitian A (1990) Spi-1 oncogene activation in Rauscher and Friend murine virus-induced acute erythroleukemias. Leukemia 4: 20–23

Morishita K, Parker DS, Mucenski ML, Jenkins NA, Copeland NG, Ihle JN (1988) Retroviral activation of a novel gene encoding a zinc finger protein in IL-3-dependent myeloid leukemia cell lines. Cell 54: 831–840

Morse B, Rotherg PG, South VJ, Spandorfer JM, Astrin SM (1988) Insertional mutagenesis of the myc locus by a LINE-1 sequence in a human breast carcinoma. Nature 333: 87–88

Mucenski ML, Taylor BA, Copeland NG, Jenkins NA (1988a) Chromosomal location of evi-1, a common site of ectropic viral integration in AKXD murine myeloid tumors. Oncogene Res 2: 219–233

Mucenski ML, Taylor BA, Ihle JN et al. (1988b) Identification of a common ecotropic viral integration site evi-1 in the DNA of AKXD murine myeloid tumors. Mol Cell Biol 8: 301–308

Mullins JI, Brody DS, Binari RCJr, Cotter SM (1984) Viral transduction of c-myc gene in naturally occurring feline leukaemias. Nature 308: 851–858

Mushinski JF, Potter M, Bauer SR, Reddy EP (1983) DNA rearrangement and altered RNA expression of the c-myb oncogene in mouse plasmacytoid lymphosarcomas. Science 220: 795–798

Neel BG, Hayward WS, Robinson HL, Fang J, Astrin SM (1981) Avian leukosis virus-induced tumors have common proviral integration sites and synthesize discrete new RNAs: oncogenesis by promoter insertion. Cell 23: 323–334

Neel BG, Gasic GP, Rogler CE et al. (1982) Molecular analysis of the c-myc locus normal tissue and in avian leukosis virus- induced lymphomas. J Virol 44: 158–166

Neil JC, Hughes D, McFarlane D et al. (1984) Transduction and rearrangement of the myc gene by feline leukemia virus in naturally occurring T-cell leukaemias. Nature 308: 814–820.

Nilsen TW, Maroney PA, Goodwin RG et al. (1985) c-erbB Activation in ALV-induced erythroblastosis: novel RNA processing and promoter insertion result in expression of an amino-truncated EGF receptor. Cell 41: 719–726

Noori-Daloii MR, Swift RA, Kung HJ, Crittenden LB, Witter RL (1981) Specific integration of REV proviruses in avian bursal lymphomas. Nature 294: 574–576

Norton PA, Coffin JM (1987) Characterization of Rous sarcoma virus sequences essential for viral gene expression. J Virol 61: 1171–1179

Nottenburg C, Varmus HE (1986) Features of the Chicken c-myc gene that influence the structure of c-myc RNA in normal cells and bursal lymphomas. Mol Cell Biol 6: 2800–2806

Nusse R (1986) The activation of cellular oncogenes by retroviral insertion. Trends Gene 2: 224–247

Nusse R, Varmus HE (1982) Many tumors induced by the mouse mammary tumor virus contain a provirus integrated in the same region of the host genome. Cell 31: 991–109

Nusse R, van Ooyen A, Rijsewijk F, van Lohuizen M, Schuuring E, van't Veer L (1985) Retroviral insertional mutagenesis in murine mammary cancer. Proc R Soc Lond 226: 3–13

Nusse R, TheunissenH, Wagenaar E, Rijsewijk F, Gennissen A, Otte A, Schuuring E, van Ooyen A (1990) The Wnt-q (int-1) oncogene promoter and its mechanism of activation by insertion of proviral DNA of the mouse mammary tumour virus. Mol Cell Biol 10: 4170–4179

O'Donnell PV, Fleissner E, Lonial H, Koehne CF, Reicin A (1985) Early clonality and high-frequency proviral integration into the c-myc locus in AKR leukemias. J Virol 55: 500–503

Paul R, Schuetze S, Kozak SL, Kabat D (1989) A common site for immortalizing proviral integrations in Friend erythroleukemia : molecular cloning and characterization. J Virol 63: 4958–4961

Payne GS, Courtneidge SA, Crittenden LB, Fadly AM, Bishop JM, Varmus HE (1981) Analysis of avian leukosis virus DNA and RNA in bursal tumors: viral gene expression is not required for maintenance of the tumor state. Cell 23: 311–322

Payne GS, Bishop JM, Varmus HE (1982) Multiple arrangements of viral DNA and an activated host oncogene in bursal lymphomas. Nature 295: 209–214

Pear WS, Nelson SF, Axelson H, Wahlström G, Bazin H, Klein G, Sümegi J (1988) Aberrant class switching juxtaposes c-myc with a middle repetitive element (LINE) and an IgH intron in two spontaneously arising rat immunocytomas.

Pecenka V, Dvorák M, Karafiát V, Sloncová E, Hlozánek I, Trávnicek M, Riman J (1988) Avian nephroblastomas induced by a retrovirus (MAV-2) lacking oncogene: II. Search for common sites of proviral integration in tumour DNA. Folia Biol (Praha) 34: 147–169

Peters G, Brookes S, Smith R, Dickson C (1983) Tumorigenesis by mouse mammary tumor virus: evidence for a common region for provirus integration in mammary tumors. Cell 33: 369–377

Peters G, Brookes S, Placzek M, Schuermann M, Michalides R, Dickson C (1989a) A putative int domain for mouse mammary tumor virus on mouse chromosome 7 is a 5' extension of int-2. J Virol 63: 1448–1450

Peters G, Brookes S, Smith R, Placzek M, Dickson C (1989b) The mouse homolog of the *hst/k-FGF* gene is adjacent to *int*-2 and is activated by proviral insertion in some virally induced mammary tumors. Proc Natl Acad Sci USA 86: 5678–5682

Pizer E, Humphries EH (1989) RAV-1 insertional mutagenesis: disruption of the c-*myb* locus and development of avian B-cell lymphomas. J Virol 63: 1630–1640

Poirier Y, Kozak C, Jolicoeur P (1988) Identification of a common helper provirus integration site in Abelson murine leukemia virus-induced lymphoma DNA. J Virol 62: 3985–3992

Quinn TP, Grandgenett DP (1988) Genetic evidence that the avian retrovirus DNA endonuclease domain of *pol* is necessary for viral integration. J Virol 62: 2307–2312

Raines MA, Lewis WG, Crittenden LB, Kung HJ (1985) c-*erb*B Activation in avian leukosis virus-induced erythroblastosis: clustered integration sites and the arrangement of provirus in the c-*erb*B alleles. Proc Natl Acad Sci USA 82: 2287–2291

Raines MA, Maihle NJ, Moscovici C, Crittenden L, Kung HJ (1988) Mechanism of c-*erb*B transduction: newly released transducing viruses retain poly(A) tracts of *erb*B transcripts and encode C-terminally intact *erb*B proteins. J Virol 62: 2437–2443

Reicin A, Yang J-Q, Marcu KB, Fleissner E, Koehne CF, O'Donnell PV (1986) Deregulation of the c-*myc* oncogene in virus-induced thymic lymphomas of AKR/J mice. Mol Cell Biol 6: 4088–4092

Robinson HL, Gagnon GC (1986) Patterns of proviral insertion and deletion in avian leukosis virus-induced lymphomas. J Virol 57: 28–36

Roelink H, Wagenaar E, Lopes da Silva S, Nusse R (1990) *Wnt*-3, a gene activated by proviral insertion in mouse mammary tumors, is homologous to *int*-1/*Wnt*-1 and is normally expressed in mouse embryos and adult brain. Proc Natl Acad Sci USA 87: 4519–4523

Rohdewohld H, Weiher H, Reik, W, Jaenisch R, Breindl M (1987) Retrovirus integration and chromatin structure: Moloney murine leukemia provial integration sites map near DNase I-hypersensitive sites. J Virol 61: 336–343

Rosson D, Dugan D, Reddy EP (1987) Aberrant splicing events that are induced by proviral integration: implications for *myb* oncogene activation. Proc Natl Acad Sci USA 84: 3137–3175

Roth MJ, Tanese N, Goff SP (1988) Gene product of Moloney leukemia virus required for proviral integration is a DNA binding protein. J Mol Biol 203: 131–139

Schnieke A, Harbers K, Jaenisch R (1983) Embryonic lethal mutation in mice induced by retrovirus insertion into the (I) collagen gene. Nature 304: 315–320.

Schubach W, Groudine M (1984) Alteration of c-*myc* chromatin structure by avian leukosis virus integration. Nature 307: 702–708

Schwartzberg PJ, Colicelli J, Goff SP (1984) Construction and analysis of deletion mutations in the *pol* gene of Moloney murine leukemia virus: a new viral function required for productive infection. Cell 37: 1043–1052

Selten G, Cuypers HT, Zijlstra M, Melief C, Berns A (1984) Involvement of c-*myc* in MuLV-induced T cell lymphomas in mice:frequency and mechanisms of infection. EMBO J 3: 3215–3222

Selten G, Cuypers HT, Berns A (1985) Proviral activation of the putative oncogene *pim*-1 MuLV induced T-cell lymphomas. EMBO J 4:1793–1798

Seperack PK, Strobel MC, Corrow DJ, Jenkins NA, Copeland NG (1988) Somatic and germ-line reverse mutation rates of the retrovirus-induced dilute coat-color mutation of DBA mice. Proc Natl Acad Sci USA 85: 189–192

Setoguchi M, Higuchi Y, Yoshida S, Nasu N, Miyazaki Y, Akizuki S, Yamamoto S (1989) Insertional activation of N-*myc* by endogeneous Moloney- like murine retrovirus sequences in macrophage cell lines derived from myeloma cell line-macrophage hybrids. Mol Cell Biol 9: 4515–4522

Sharp PA, Marciniak RA (1989) HIV TAR: an RNA enhancer? Cell 59: 229–230

Shaw G, Kamen R (1986) A conserved AU sequence from the 3' untranslated region of GM-CSF mRNA mediates selective mRNA degradation. Cell 46: 659–667

Shen-Ong GLC, Potter M, Mushinski JF, Lavu S, Reddy EP (1984) Activation of the c-*myb* locus by viral insertional mutagenesis in plasmacytoid lymphosarcomas. Science 226: 1077–1080

Shen-Ong GLC, Morse, HC III, Potter M, Mushinski JF (1986) Two modes of c-*myb* activation in virus-induced mouse myeloid tumors. Mol Cell Biol 6: 380–392

Shih C-C, Stoye JP, Coffin JM (1988) Highly preferred targets for retrovirus integration. Cell 53: 531–537

Shih C-K, Linial M, Goodenow, MM, Hayward WS (1984) Nucleotide sequence 5' of the chicken c-*myc* coding region. Localization of a noncoding exon that is absent from *myc* transcripts in most avian leukosis virus-induced lymphomas. Proc Natl Acad Sci USA 81: 4697–4701

Shimotono K, Mizutani S, Temin HM (1980) Sequence of retrovirus provirus resembles that of bacterial transposable elements. Nature 285: 550–554

Silver J, Kozak C (1986) Common proviral integration region on mouse chromosome 7 in lymphomas and myelogenous leukemias induced by Friend murine leukemia virus. J Virol 57: 526–533

Simon MC, Smith RE, Hayward WS (1984) Mechanisms of oncogenesis by subgroup F avian leukosis viruses. J Virol 52: 1–8

Simon MC, Neckameyer WS, Hayward WS, Smith RE (1987) Genetic determinants of neoplastic diseases induced by a subgroup F avian leukosis virus. J Virol 61: 1203–1212

Sola B, Simon D, Mattei M-G et al. (1988) Fim-1, Fim- 2/c-fms, and Fim-3, three common integration sites of Friend murine leukemia virus in myeloblastic leukemias, map to mouse chromosomes 13, 18, and 3, respectively. J Virol 52: 1–8

Soret J, Dambrine G, Perbal B (1989) Induction of nephroblastoma by myeloblastosis-associated virus type 1: state of proviral DNAs in tumor cells. J Virol 63: 1803–1807

Southgate C, Zapp ML, Green MR (1990) Activation of transcription by HIV-1 tat protein tethered to nascent RNA through another protein. Nature 345: 640–642

Spence SE, Gilbert DJ, Swing DA, Copeland NG, Jenkins NA (1989) Spontaneous germ line virus infection and retroviral insertional mutagenesis in eighteen transgenic Srev lines of mice. Mol Cell Biol 9: 177–184

Stavenhagen JB, Robins DM (1988) An ancient provirus has imposed androgen regulation on the adjacent mouse sex-limited protein gene. Cell 55:247–254

Steffen D (1984) Proviruses are adjacent to c-myc in some murine leukemia virus-induced lymphomas. Proc Natl Acad Sci USA 81: 2097–2101

Stewart MA, Forrest D, McFarlane R, Onions D, Wilkie N, Neil JC (1986) Conservation of the c-myc coding sequence in transduced feline v-myc genes. Virology 154: 121–134

Stocking C, Loliger C, Kawai M, Suciu S, Gough N, Osertag W (1988) Identification of genes involved in growth autonomy of hematopoietic cells by analysis of factor-independent mutants. Cell 53: 869–879

Stoye JP, Fenner S, Greenoak GE, Moran C, Coffin JM (1988) Role of endogenous retroviruses as mutagens: the hairless mutation of mice. Cell 54: 383–391

Swanstrom R, Parker RC, Varmus HE, Bishop JM (1983) Transduction of a cellular oncogene: the genesis of Rous sarcoma virus. Proc Natl Acad Sci USA 80: 2519–2523

Swift RA, Shaller E, Witter RL, Kung HJ (1985) Insertional activation of c-myc by reticuloendotheliosis virus in chicken B lymphoma: nonrandom distribution and orientation of the proviurses. J Virol 54: 869–872

Swift RA, Boerkoel C, Ridgway A, Fujita DJ, Dodgson JB, Kung HJ (1987) B-lymphoma induction by reticuloendotheliosis virus: characterization of a mutated chicken syncytical virus provirus involved in c-myc activation. J Virol 61: 2084–2090

Taketo M, Shaffer DJ, (1989) Deletions in a recombinant retrovirus genome associated with its expression in embryonal carcinoma cells. J Virol 63: 4431–4433

Terry R Solits DA, Katzman M, Cobrinik D, Leis J, Skalka AM (1988) Properties of avian sarcoma-leukosis virus pp 32-related pol-endonucleases produced in E coli. J Virol 62: 2358–2365

Trusko SP, Hoffman EK, George DL (1989) Transcriptional activation of cKi-nas proto-oncogene resulting from retroviral promoter insertion. Nucl Acid Res 17: 9259–9265

Tsichlis PN,Strauss PG, Fu Hu L (1983) A common region for proviral DNA integration in MoMuLV-induced rat thymic lymphomas. Nature 302: 445–449

Tsichlis PN, Strauss PG, Lohse MA (1985) Concerted DNA rearrangements in Moloney murine leukemia virus-induced thymomas: a potential synergistic relationship in oncogenesis. J Virol 56: 258–267

Tsichlis PN, Shepherd BM, Bear SE (1989) Activation of the Mlvi-1/mis-1/prt-1 locus in Moloney murine leukemia virus-induced T-cell lymphomas. Proc Natl Acad Sci USA 86: 5487–5491

Tsichlis PN, Lee JS, Bear SE et al. (1990) Activation of multiple genes by provirus integration in the Mlvi-4 locus in T-cell lymphomas induced by Moloney murine leukemia virus. J Virol 64: 2236–2244

Tsukamoto AS, Grosschedl R, Guzman RC, Parslow T, Varmus HE (1988) Expression of the int-1 gene in transgenic mice is associated with mammary gland hyperplasia and adenocarcinomas in male and female mice. Cell 55: 619–2625

van Lohuizen M, Verbeek S, Krimpenfort P et al. (1989) Predisposition to lymphomagenesis in pim-1 transgenic mice: cooperation with c-myc and N-myc in murine leukemia virus-induced tumors. Cell 56: 673–682

Varmus H, Brown P (1989) Retroviruses. I: Berg DE, Howe MM Mobile DNA American Society for Microbiology, Washington DC, pp 53–108

Varmus HE, Quintrell N, Ortiz S (1981) Retroviruses as mutagens: insertion and excision of a nontransforming provirus alter expression of a resident transforming provirus. Cell 25: 23–26

Vijaya S, Steffen DL, Robinson HL (1986) Acceptor sites for retroviral integrations map near DNase-I hypersensitive sites in chromatin. J Virol 60: 683–692

Vijaya S, Steffen DL, Kozak C, Robinson HL (1987) *Dsi*-1 a region with frequent proviral insertions in Moloney murine leukemia virus-induced rat thymomas. J Virol 61: 1164–1170

Villemur R, Monczak Y, Rassart E, Kozak C, Jolicoeur P (1987) Identification of a new common provirus integration site in Gross passage A murine leukemia virus-induced mouse thymoma DNA. Mol Cell Biol 7: 512–522

Villeneuve L, Rassart E, Jolicoeur P, Graham M, Adams JM (1986) Proviral integration site *mis-1* in rat thymomas corresponds to the *pvt-1* translocation breakpoint in murine plasmacytomas. Mol Cell Biol 6: 1834–1837

Voronova AF, Sefton BM (1986) Expression of a new tyrosine protein kinase is stimulated by retrovirus promoter insertion. Nature 319: 682–685

Weinstein Y, Cleveland JL, Askew DS, Rapp UR, Ihle JN (1987) Insertion and truncation of c-*myb* by murine leukemia virus in a myeloid cell line derived from, cultures of normal hematopoietic cells. J Virol 61: 2339–2343

Westaway D, Payne G, Varmus HE (1984) Proviral deletions and oncogene base- substitutions in insertionally mutagenized c-*myc* alleles may contribute to the progression of avian bursal tumors. Proc Natl Acad Sci USA 81: 843–847

Westaway D, Papkoff J, Moscovici C, Varmus HE (1986) Identification of a provirally activated c-ha-*ras* oncogene in an avian nephroblastoma via a novel procedure: cDNA cloning of a chimaeric viral-host transcript. EMBO J 5: 301–309.

Wolf D, Rotter V (1984) Inactivation of p53 gene expression by an insertion of Moloney murine leukemia virus-like DNA sequences. Mol Cell Biol 4: 1402–1410

Wolf D, Harris N, Rotter V (1984) Reconstitution of p53 expression in a nonproducer Ab-MuLV-transformed cell line by transfection of a functional p53 gene. Cell 38: 119–126

Mutagenesis of the Mouse Germline Using Retroviruses*

L. F. Lock, N. A. Jenkins, and N. G. Copeland

1   Introduction 27
1.1 Mutations Produced by Infection of Embryos 28
1.2 Mutations Produced by Infection of ES Cells 29
1.3 Mutations that Have Arisen Spontaneously 30

2   The Mechanism of Spontaneous Acquisition of Germline Proviruses 32
2.1 The Acquisition of New Germline Proviruses in SWR/J–RF/J Hybrid Mice 33
2.2 Development of a Novel Method to Infect the Mouse Germline 37
2.3 Mutations Identified in SWR/J–RF/J Hybrid Mice 37

3   Concluding Remarks 38

References 39

1 Introduction

Traditionally, mutations in the mouse have been identified as spontaneously occurring mutations in feral or laboratory-maintained mice or as induced mutations in radiation-or chemical-treated mice (GREEN 1981). Although much has been learned from the study of these mutations, molecular analysis of the mutated genes has often been hampered by the lack of molecular access to them. To circumvent this problem, transposable elements can be used as insertional mutagens. The presence of a transposable element within or near the mutant gene acts as a molecular tag which aids in cloning the mutated gene of interest. This approach has been used with great success in several organisms, most notably *Drosophila melanogaster* (RUBIN 1988).

In the mouse, retroviruses represent a transposable element that can be used as an insertional mutagen and, as such, they have several advantages. First, retroviruses integrate into the host cell genome as a single integrated retrovirus (referred to as a provirus) without significant rearrangement of host cell DNA, thereby creating an ideal molecular tag (WEISS et al. 1984). Second,

* L.F.L. is supported by a Damon Runyon-Walter Winchell Cancer Fund Fellowship, DRG-918. This work was sponsored by the National Cancer Institute, DHHS, under contract N01-CO-74101 with ABL.

Mammalian Genetics Laboratory, ABL-Basic Research Program, NCI-Frederick Cancer Research and Development Center, Frederick, MD 21702, USA

the sites of retroviral integration appear to be relatively nonspecific with respect to the host cell DNA, although a bias may exist for integration near actively transcribed genes and preferred sites may also exist (WEISS et al. 1984; VIJAYA et al. 1986; ROHDEWOHLD et al. 1987; SHIH et al. 1988). Third, the provirus can cause a mutant phenotype by affecting gene function when integrated into either coding or noncoding sequence (SCHNIEKE et al. 1983; HARBERS et al. 1984; HUTCHISON et al. 1984; STROBEL et al. 1990). This increases the proportion of the genome into which proviral integration will induce a mutant phenotype.

Several methods can be employed to use retroviruses as insertional mutagens in the mouse. First, they can be introduced into the mouse germline experimentally by infection of embryos, either in vitro or in utero (JÄHNER and JAENISCH 1980; JAENISCH 1980). Second, embryonic stem (ES) cells can be infected in vitro and incorporated into the germline of a chimeric mouse (KUEHN et al. 1987). Third, new proviruses can, on rare occasions, spontaneously integrate into the mouse germline (ROWE and KOZAK 1980; LANGDON et al. 1984).

1.1 Mutations Produced by Infection of Embryos

To experimentally induce mutations via embryo infection, mouse embryos can be infected at two different stages of embryonic development. Four- to eight-cell embryos can be isolated, treated with protease to remove the zona pellucida (which appears to be an effective barrier to virus infection), and cultured in vitro, either in medium to which an infectious virus stock has been added or in the presence of cells that produce infectious virus (JÄHNER and JAENISCH 1980). The embryos develop in vitro to the blastocyst stage and are then transferred to the uteri of pseudopregnant foster female mice. Alternatively, midgestation embyros can be infected in utero by injection of virus directly into the embryo at day 8–9 of gestation (JAENISCH 1980). Mice produced using either infection method are mated and their progeny analyzed for the presence of new endogenous proviruses by Southern blot analysis. New germline proviruses are observed in the progeny of mice infected at preimplantation at a much higher frequency than those infected at midgestation (JAENISCH 1980; SORIANO ot al. 1987). Four mutations induced by infection of embryos have been reported; three are recessive lethal mutations and one is a kidney defect (JAENISCH 1988). Two of the recessive lethal mutations, *Mov-13* and *Mov-34*, have been further characterized both phenotypically and molecularly (Table 1).

The *Mov-13* mutation originated from a postimplantation embryo that had been infected in utero with Moloney murine leukemia virus (M-MuLV; JAENISCH 1980). Mice homozygous for the *Mov-13* integration arrest in development at day 11–12 and die at day 13–14 of gestation (JAENISCH et al. 1983). Erythropoietic and mesenchymal cell necrosis is evident prior to death, which results from rupture of one of the major blood vessels (LOHLER et al. 1984). Northern analysis reveals that the *Mov-13* gene is transcribed in normal embryos beginning at about day 12 of gestation (SCHNIEKE et al. 1983). High expression is found in

Table 1. Retrovirally induced mutations in the mouse

Mutation	Origin	Phenotype	Gene affected
Mov-13	M-MuLV infection of midgestation embryo	Lethal at days 12–14	α1(I) Collagen
Mov-34	M-MuLV supF infection of preimplantation embryo	Lethal just after implantation	Unknown
d^v	Spontaneous	Lightened coat color	Unknown
hr	Spontaneous	Hairless	Unknown
Srev-5	Spontaneous acquisition in SWR/J–RF/J mice	Lethal at days 8.5–11.5	Unknown

M-MuLV, Moloney murine leukernia virus

some mesoderm-derived cells, such as fibroblasts and myogenic cells, whereas no expression is found in endoderm-, ectoderm-, and other mesoderm-derived cells such as hemopoietic cells (SCHNIEKE et al. 1983). This pattern of expression eventually led to the determination that the *Mov-13* provirus is integrated in the first intron of the α1(I) collagen gene (SCHNIEKE et al. 1983; HARBERS et al. 1984). Analysis of *Mov-13* homozygous embryos suggests that normal transcription of the α1(I) collagen gene is blocked by the *Mov-13* integration (SCHNIEKE et al. 1983; HARBERS et al. 1984). Studies of the *Mov-13* mutation indicate that normal transcription of the α1(I) collagen gene is required for normal development by day 11 of gestation.

The *Mov-34* mutation was derived by infection of preimplantation embryos with M-MuLV containing a bacterial suppressor tRNA gene in each LTR (SORIANO et al. 1987). When mice heterozygous at *Mov-34* were intercrossed, homozygous *Mov-34* progeny were not born, suggesting that the *Mov-34* is a recessive lethal mutation. Examination of embryos from *Mov-34* heterozygous intercross matings revealed that homozygous *Mov-34* embryos are viable prior to implantation, abnormal at day 6 of gestation, and absent at day 8 of gestation. Using a DNA probe that flanks the *Mov-34* integration site, Northern analysis demonstrated a single transcript which is abundantly expressed in embryos at day 10 of gestation and in many adult tissues. The transcript is reduced in abundance in heterozygous *Mov-34* mice, suggesting that the *Mov-34* provirus interferes with gene transcription. Presumably, the reduction in transcription of this gene results in lethality in homozygous *Mov-34* embryos. This suggests that a normal level of the gene affected by the *Mov-34* integration is required in embryonic development just after implantation.

1.2 Mutations Produced by Infection of ES Cells

New proviruses can be introduced into the mouse germline by infection of ES cells in vitro followed by the incorporation of the infected cells into the germline of chimeric mice. ES cells derive from the pluripotent stem cells of the peri-implantation embryo (EVANS and KAUFMAN 1981; MARTIN 1981). Some established ES cell lines retain pluripotency and are capable of contributing to all tissues of a chimeric mouse, including the germline, when reintroduced into the embryo (BRADLEY et al. 1984). Such ES cells are infected in vitro either by addition of an infectious virus stock to the culture medium or by cocultivation with cells that produce infectious virus. Infected cells can then be selected in vitro, if desired, and introduced into the germline by injection of the ES cells into the cavity of an embryo at the blastocyst stage of development. The chimeric embryos are transferred to the uteri of pseudopregnant foster female mice. The progeny of the chimeric mice which derive from the ES cells are then analyzed for mutant phenotypes associated with the presence of a provirus.

Mice with provirally induced mutations in the X-linked gene *hprt* have been produced by infecting ES cells with retroviruses, selecting for HPRT-deficient cells, and incorporating these cells into the germline of a chimeric mouse by blastocyst injection (Table 1; KUEHN et al. 1987). Females heterozygous for the *hprt* mutation, as expected because of random X-chromosome inactivation, had two populations of cells, those with normal HPRT activity and those with no HPRT activity. When these females were mated to normal males, female and male progeny were produced at approximately equal frequencies, suggesting that the *hprt* mutation was not lethal in hemizygous males. Lack of HPRT activity was demonstrated in males hemizygous for the *hprt* mutation. However, despite the lack of HPRT activity, these mice were phenotypically normal. A completely normal phenotype was not predicted, based on the manifestations of HPRT deficiency associated with the human Lesch-Nyhan syndrome, in which affected males have metabolic, neurological, and behavioral symptoms. The lack of a mutant phenotype with the loss of HPRT activity indicates that *hprt* gene function is not essential in the mouse. The circumstances that account for the difference in phenotype between mouse and human are unclear, but these results demonstrate that retroviral infection and selection of ES cells followed by incorporation into the germline of chimeric mice is a viable method for producing insertional mutations in mice.

1.3 Mutations that Have Arisen Spontaneously

Endogenous type C murine leukemia proviruses are present in the mouse germline in multiple nonallelic sites. They consist of either all or part of an integrated retroviral genome including three genes, *gag*, *pol*, and *env*, flanked by two long terminal repeats (LTRs). Endogenous MuLVs fall into three categories based on host range: ecotropic viruses that replicate in mouse cells, xenotropic viruses that replicate in nonmouse cells, and polytropic viruses that replicate in either mouse or nonmouse cells. Host range is conferred by the *env* gene, which encodes a protein that interacts with a cell encoded receptor during infection. In the mouse, the number of xenotropic and polytropic endogenous proviral loci is so large that it has been difficult to study individual proviral loci of these classes. However, recent identification of divergent regions of these endogenous proviruses has enabled development of DNA probes that recognize smaller, more workable subsets of these loci (STOYE and COFFIN 1987, 1988). In contrast, relatively few endogenous ecotropic murine leukemia proviruses (or *Emv* loci) are present in the mouse germline. All inbred mouse strains that have been examined contain between 0 and 11 *Emv* loci in their genome (JENKINS et al. 1982; TAYLOR and ROWE 1989a). With respect to the expression of *Emv* loci, the inbred mouse strains fall into three categories: nonviremic, poorly viremic, and highly viremic. Although there is a strong correlation between the number of *Emv* loci present and the amount of virus production, this correlation is not absolute (CHATTOPADHYAY et al. 1980; JENKINS et al. 1982). The expression of *Emv* loci is affected by many factors acting either in cis or trans. The rare acquisition of new germline ecotropic proviruses is usually observed in inbred strains that have a complete nondefective *Emv* locus and are highly viremic.

Mutations at the *dilute* and *hairless* genes of the mouse are causally associated with endogenous proviruses (Table 1; JENKINS et al. 1981; STOYE et al. 1988). Presumably, the endogenous proviruses that cause the mutations arose by spontaneous infection of the germline of the mice in which the mutations occurred. Many other existing mouse mutations may also be caused by the integration of endogenous proviruses. Elucidation of such causative relationships awaits detailed characterization of the numerous and varied endogenous proviruses present both in wild-type and mutant mice.

Over 200 alleles of the *dilute* gene, which causes lightening of coat color, have been identified in the mouse. In many cases, these mutations also cause neurological abnormalities and lethality. The original dilute mutation, d^v, is an spontaneous mutation that causes only a lightening of the coat color. This change in coat color does not result from a reduction in melanin synthesis, but rather an altered melanocyte morphology (SILVERS 1979). As compared to the wild type, the melanocyte dendritic processes of homozygous d^v mice are reduced in size and number. Since the dendrites of melanocytes function in the deposition of pigment in to the hair bulb, pigmentation of the hair is reduced in homozygous d^v mice. By Southern blot analysis using an ecotropic virus-specific probe, coincidence was found between the d^v mutation and an endogenous ecotropic murine leukemia provirus, *Emv-3* (JENKINS et al. 1981). The spontaneous reversion rate of the d^v mutation of wild-type coat color is unusually high compared to other recessive coat color mutations (SEPERACK et al. 1988). Analysis of the ecotropic proviral content of these revertants indicated that reversion of the d^v allele to wild-type coat color is accompanied by the loss of most of *Emv-3* (COPELAND et al. 1983; SEPERACK et al. 1988). Because a single LTR was retained in revertant mice, the loss of *Emv-3* probably resulted from homologous recombination between the viral LTRs. *Emv-3* was used as a molecular tag to clone genomic DNA from the d^v mutation (RINCHIK et al. 1986). Molecular analysis of the DNA flanking *Emv-3* suggests that, in the d^v mutation, the provirus is integrated into an intron of the *dilute* gene

(HUTCHISON et al. 1984; STROBEL et al. 1990; P.K. SEPERACK, M.C. STROBEL, N.G. COPELAND, N.A. JENKINS, manuscript in preparation). Several transcripts of the dilute gene were identified by Northern analysis of RNA from a B16 melanoma cell line, which is wild-type at the *dilute* locus (JENKINS et al. 1989; P.K. SEPERACK, M.C. STROBEL, N.G. COPELAND, N.A. JENKINS, manuscript in preparation). Several cDNA clones have been isolated from a library prepared from B16 melanoma cell RNA. When these cDNA clones were used in Northern analysis, transcripts were observed in several adult tissues including brain. Further examination of the *dilute* gene and its products, combined with genetic and phenotypic analysis of the numerous existing dilute mutations (RUSSELL 1971), should reveal the nature and function of the *dilute* gene.

The hairless mutation, *hr*, is a spontaneous autosomal recessive mutation that maps to mouse chromosome 14 (GREEN 1981). Its effects are pleiotropic; mice homozygous for the *hr* mutation are hairless at 3–4 weeks of age, lack mammary gland function, have abnormal differentiation of thymocytes, and have a high incidence of lymphoid leukemia and UV- and carcinogen-induced skin neoplasms (MEIER et al. 1969; MORRISSEY et al. 1980; GALLAGHER et al. 1984; POLAND et al. 1984). By Southern blot analysis of genomic DNA using oligo-nucleotide probes specific to subsets of the endogenous polytropic proviruses, coincidence was discovered between the *hr* mutation and an endogenous polytropic provirus, MX40 (STOYE and COFFIN 1988; STOYE et al. 1988). Mice carrying the *hr* mutation have an extra provirus as compared to wild-type mice. Proof that the provirus caused the *hr* mutation was provided by analysis of a spontaneous revertant of the hairless mutation (STOYE et al. 1988). Reversion to a wild-type phenotype was accompanied by loss of most of the MX40 provirus; a single LTR was retained in the revertant, suggesting that the provirus was lost by homologous recombination between the proviral LTRs. Reversion to a normal phenotype despite retention of the LTR, along with sequence analysis of flanking cellular DNA, suggests that the MX40 provirus is not located in the coding region of the *hairless* gene. The precise structural and functional relationship between the MX40 provirus and the *hairless* gene remains to be determined.

2 The Mechanism of Spontaneous Acquisition of Germline Proviruses

Ideally, the efficient use of retroviruses as insertional mutagens requires a technically simple method to introduce new proviruses into the mouse germline at high frequency. Infection of embryos or ES cells are relatively labor intensive and technically difficult, requiring sophisticated manipulations of embryos and/or cells in culture. In contrast, spontaneous acquisition of new germline proviruses is technically straightfoward; however, it occurs too infrequently in most inbred strains to be an efficient method for introducing new proviruses into the mouse germline. The development of mouse strains that spontaneously acquire new germline proviruses at high frequency would provide a technically simple system for retroviral insertional mutagenesis. Two such mouse strains have been identified. MEV/1Ty mice have 11 ecotropic MuLVs in their genome (TAYLOR and ROWE 1989b). New germline proviruses are acquired in these mice at a frequency of about 0.05 new proviruses per mouse (TAYLOR and ROWE 1989a). Similarly, SWR/J–RF/J hybrid mice acquire new germline proviruses more frequently than previously reported for other mice (JENKINS and COPELAND 1985; BAUTCH 1986). The frequency of acquisition of germline ecotropic proviruses in these hybrid mice ranges from about 0.1 to 0.5 new proviruses per mouse. The high frequency of proviral acquisition observed in SWR/J–RF/J mice provides an opportunity to design experiments to elucidate the mechanism by which new *Emv* loci are acquired. By understanding the mechanism by which *Emv* loci are acquired, it may be possible to increase the proviral acquisition frequency even higher facilitating the use of this system for insertional mutagenesis.

2.1 The Acquisition of New Germline Proviruses in SWR/J–RF/J Hybrid Mice

SWR/J–RF/J hybrid mice are produced by mating mice of two inbred mouse strains, SWR/J and RF/J. SWR/J mice have no *Emv* loci (JENKINS et al. 1982). In contrast, RF/J mice carry three *Emv* loci: *Emv-1*, *Emv-16*, and *Emv-17* (JENKINS et al. 1982). *Emv-1* is a defective provirus located on mouse chromosome 5 (MCCUBREY et al. 1982). *Emv-16* and *Emv-17* are closely linked on mouse chromosome 1 but are not tandemly duplicated (JENKINS and COPELAND 1985). *Emv-16* and *Emv-17* are non-defective proviruses capable of generative infectious virus in vitro (L.F. LOCK, A.M. BUCHBERG, E. KESHET, N.G. COPELAND, N.A. JENKINS, manuscript in preparation). When *Emv-16* and *Emv-17* are backcrossed onto the SWR/J strain, new germline proviruses are observed in the progeny of N_2 females that carry *Emv-16* and *Emv-17* (JENKINS and COPELAND 1985; BAUTCH 1986; PANTHIER and CONDAMINE 1987). As many as 75% of some liters carry new germline ecotropic proviruses. A single mouse can carry as many as ten new proviruses. The newly acquired proviruses are present in the founder mice in less than one copy per cell. The extent of mosaicism in somatic tissues, which ranges from 0.04–0.74 copies per diploid genome, suggests that viral integration occurs after DNA replication in the zygote but prior to DNA replication in the four-cell embryo (BAUTCH 1986; SPENCE et al. 1989). Furthermore, the similarity between the extent of mosaicism in the somatic tissues and the frequency of germline transmission suggests that viral integration occurs prior to the delineation of the somatic and germinal lineages in the embryo (BAUTCH 1986; SPENCE et al. 1989). Newly acquired SWR/J–RF/J ecotropic proviral loci, referred to as *Srev* loci, have been extensively characterized with respect to their genome structure and expression (SPENCE et al.

1989). Analysis of restriction endonuclease sites of *Srev* loci suggests that about 78% appear to be normal, complete proviruses. Of those that appear complete, 36% are expressed leading to the production of infectious virus.

The new germline proviruses present in SWR/J–RF/J hybrid mice could arise by extracellular virus infection, intracellular retroposition, DNA-mediated transposition, or other unknown mechanisms. Ecotropic virus production in the maternal environment appears to play a major role in the acquisition of proviruses in the progeny (JENKINS and COPELAND 1985). New proviruses are only observed in the progeny of SWR/J–RF/J hybrid females that carry *Emv-16* and *Emv-17*. Virus production in these mice is affected by two factors: maternal resistance factor (MRF), a nongenetic factor conferred on the progeny by the mother, and *Fv-1*, the major locus in the mouse that controls ecotropic virus production (PINCUS et al. 1971; MAYER et al. 1978; CHEN et al. 1980; MAYER et al. 1980). RF/J mice are restricted for virus production; they have MRF and carry a nonpermissive allele at the *Fv-1* locus, $Fv-1^{nr}$. In contrast, SWR/J mice are permissive for virus production; they do not have MRF and carry a permissive allele at the *Fv-1* locus, $Fv-1^{n}$. Because of the mating protocol used to backcross *Emv-16* and *Emv-17* onto the SWR/J strain background, the female mice whose progeny have new germline proviruses are permissive for virus production; they do not have the MRF and are permissive at the *Fv-1* locus. When high-titer virus production is blocked by altering the mating protocol to retain either a restrictive allele at the *Fv-1* locus or the MRF, high-frequency acquisition of new germline proviruses is not observed in the progeny (JENKINS and COPELAND 1985). Also, in mice that have new *Srev* loci, additional new germline proviruses are only observed in the progeny of mice that carry *Srev* loci capable of producing infectious virus (SPENCE et al. 1989). In accordance with these observations, the new germline ecotropic proviruses that are occasionally observed in other highly viremic mice are only observed in the progeny of virus-positive females (ROWE and KOZAK 1980).

The observed correlation between high-titer virus production in the mother and high-frequency acquisition of new proviruses in the progeny suggests that the new proviruses are acquired by an infection mechanism. To test this hypothesis, an ovarian transplantation study was performed in which ovaries from mice that have no endogenous ecotropic proviruses were transplanted to viremic hosts that carry *Emv-16* and *Emv-17* and mated to males that have no endogenous ecotropic proviruses (Fig. 1; LOCK et al. 1988). The only ecotropic proviruses present in this experiment are *Emv-16* and *Emv-17*, present in the host females. New germline proviruses were observed in the progeny of the donor ovary. Since the donor ovaries have no *Emv* loci, virus produced by the host must have infected cells within or derived from the donor ovary to produce new germline proviruses in donor ovary derived progeny. This result provides proof that new germline proviruses can be acquired by an infection mechanism.

It is possible that germline infection could occur in any or all cells of the female germline, including primordial germ cells, oogonia, and oocytes. In addition, infection occurring after fertilization in the cells of the early embryo

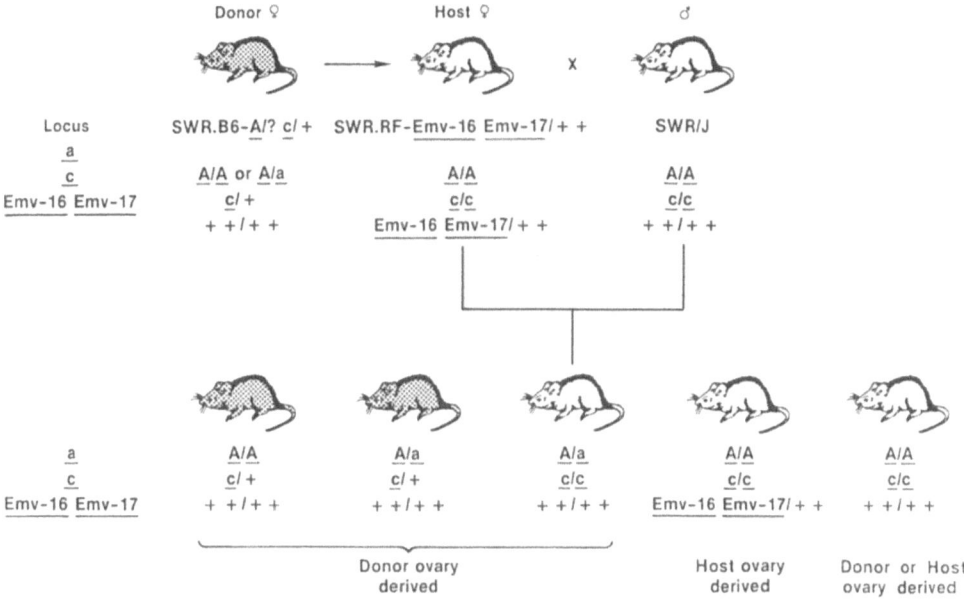

Fig. 1. Ovarian transplantations: genotypes and phenotypes of SWR.B6-*A/?, c/ +*, SWR.RF-*Emv-16 Emv-17/ + +*, and SWR/J parental mice and their progeny. Ovaries from SWR.B6-*A/?,c/ +* mice were transplanted to SWR.RF-*Emv-16 Emv-17/ + +* hosts. Ovarian transplantation recipients were mated to SWR/J male mice. Five classes of progeny were distinguished by coat color, genotype at the *agouti* locus, and the presence or absence of *Emv-16* and *Emv-17*. *Stippled mice* represent mice with agouti coat color; *unstippled mice* represent mice with albino coat color. (Reprinted from LOCK et al. 1988)

could give rise to progeny carrying new germline proviruses. Three observations suggest that the oocyte is the primary target of infection. First, the results of the ovarian transplantation study already described suggest that the oocyte or early embryo is suceptible infection. The only germ cells present in the donor ovaries were first meiotic prophase-arrested oocytes and cells derived from these oocytes. Since donor ovary-derived progeny acquired new germline proviruses, the developmental stages suceptible to infection must include the oocyte prior to, during, or after ovulation and the cells of the early embryo. Second, in situ hybridization studies of ecotropic virus-specific RNA in the genital tract of SWR/J females carrying *Emv-16* and *Emv-17* suggests that there is a potential source of infectious virus in the vicinity of both the oocyte and early embryo (PANTHIER and CONDAMINE 1987; LOCK et al. 1988). Third, the results of a second ovarian transplantation experiment suggest that the oocyte is susceptible to infection prior to, during, or just after ovulation. In this experiment, ovaries from SWR/J mice carrying *Emv-16* and *Emv-17* were transplanted to C57BL/6J nude hosts and then mated to SWR/J males (Fig. 2; LOCK et al. 1988). C57BL/6J nude hosts were chosen for several reasons including the fact that these mice carry a restrictive allele at the *Fv-1* locus which should suppress

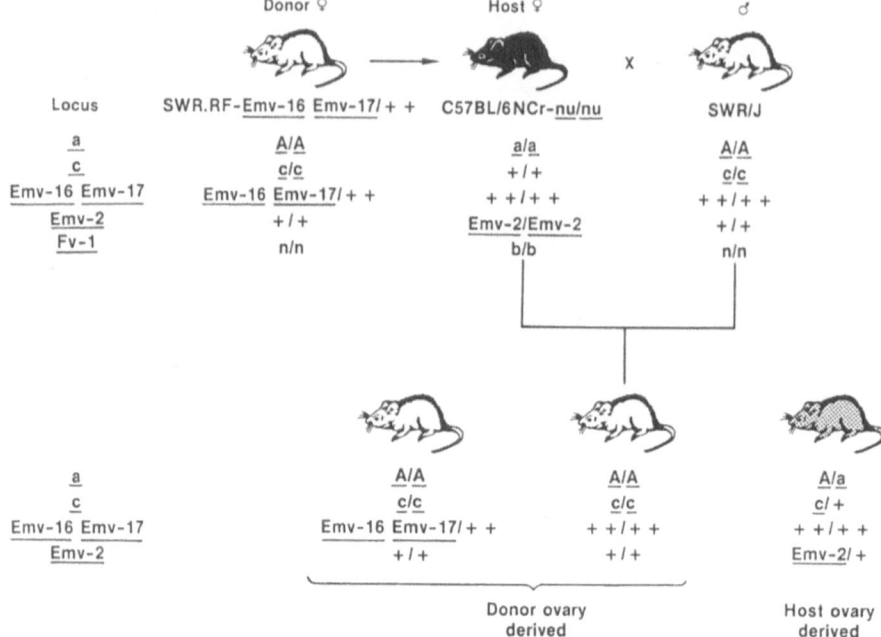

Fig. 2. Ovarian transplantations: genotypes and phenotypes of SWR.RF-*Emv*-16, *Emv*-17, C57BL/6NCr-*nu/nu*, and SWR/J parental mice and their progeny. Donor ovaries were obtained from SWR.RF-*Em*16, *Emv*-17 mice. The host mice were C57BL/6NCr-*nu/nu*. After ovarian transplantation the mice were mated to SWR/J male mice. Three classes of progeny were distinguished by coat color and the presence or absence of *Emv*-16, *Emv*-16, and *Emv*-2. *Shaded mice* represent mice with black coat color; *stippled mice* represent mice with agouti coat color; *unshaded mice* represent mice with albino coat color. (Reprinted from Lock et al. 1988)

tho cproad of virus from the donor ovary to the host. New germline proviruses were observed in donor ovary-derived progeny. Since both the host and male mice carry no replication-competent *Emv* loci and virus spread is restricted in the host, the donor ovary is the only source of infectious virus in this experiment. Infection must have occurred within or very near the donor ovary to produce progeny from the donor ovary with new germline proviruses. This suggests that the oocyte prior to, during, or just after ovulation is susceptible to infection. Comparison of the frequency of proviral acquisition in donor ovary-derived progeny to that observed in untreated SWR/J–RF/J progeny (0.15 vs 0.10 new proviruses per mouse) suggests that the oocyte is the primary target of infection. However, it is possible that some of the new germline proviruses arise by infection of other target cells including the cells of the early embyro.

2.2 Development of a Novel Method to Infect the Mouse Germline

In female SWR/J mice that carry *Emv-16* and *Emv-17*, expression of the endogenous proviruses leads to the production of virus which infects the oocyte either before, during, or just after ovulation. Since integration of the virus DNA into the cellular DNA appears to occur preferentially after entry into the DNA synthetic phase of the cell cycle (S phase; VARMUS et al. 1973), viral DNA integration would tend to occur when the cell cycle resumes after fertilization. Such a mechanism of oocyte infection, followed by integration early in embryonic development, would account for the mosaicism observed in founder mice (0.04–0.74 copies per diploid genome; SPENCE et al. 1989). Assuming random allocation of cells into embryonic and extraembryonic cell lineages, viral integration between the S phases that occur in the one- and two- cell embryo would give rise to mice carrying new proviruses present at about 0.5 copies per diploid genome. Integration at later developmental stages, non-random allocation of cells to embryonic and extraembryonic lineages, or both would result in proviruses present in founder mice at more or less than 0.5 copies per diploid. genome.

The proposed mechanism of proviral acquisition in SWR/J–RF/J mice leads to the prediction that new germline ecotropic proviruses should arise in SWR/J females induced to become viremic by exogenous administration of infectious virus derived from *Emv-16* and *Emv-17*. This prediction was tested by infecting SWR/J female mice, as newborns, with virus from *Emv-16* and *Emv-17* (LOCK et al. 1988; PANTHIER et al. 1988). At weaning these mice were viremic and had a pattern of ecotropic viral RNA expression in the genital tract that was indisting-uishable from that observed in SWR/J–RF/J females that carry *Emv-16* and *Emv-17*. New germline proviruses were observed in the progeny of these mice, demonstrating that the germline of SWR/J females can be infected by exogenously administered virus derived from *Emv-16* and *Emv-17*. The ability to infect the germline by exogenous virus infection constitutes a novel, extremely simple, efficient method to introduce DNA into the mouse germline.

2.3 Mutations Identified in SWR/J–RF/J Hybrid Mice

To date, 22 independent lines of mice have been established, each carrying a single *Srev* locus (SPENCE et al. 1989, S.E. SPENCE et al., unpublished results). During the propagation of these lines, no dominant mutations were evident. Similarly, no visible recessive or fertility mutations were observed. In contrast, 1 of the 22 *Srev* proviruses examined is associated with prenatal lethality (SPENCE et al. 1989; S.E. SPENCE, N.G. COPELAND, N.A. JENKINS, manuscript in preparation). When heterozygous carriers of the *Srev-5* provirus were intercrossed, no homo-zygous offspring were observed indicating that *Srev-5* is a recessive prenatal lethal mutation. The proportion of *Srev* loci that cause a mutant phenotype, approximately 5%, is similar to that observed using other methods to introduce retroviruses into the mouse germline (GRIDLEY et al. 1987; JAENISCH 1988). To determine the time at which *Srev-5* homozygote embryos die, Southern analysis of genomic DNA obtained from midgestation embryos of intercross matings was carried out using a probe from mouse cellular DNA that flanks the *Srev-5* integration site (S.E. SPENCE, N.G. COPELAND, N.A. JENKINS, manuscript in preparation). *Srev-5* homozygotes were not present at day 11.5 and 13.5 of gestation. In contrast, they were present at the expected frequency at day 8.5 of gestation and in reduced frequencies at days 9.5 and 10.5. This suggests that the lethality associated with the *Srev-5* mutation occurs between 8.5 and 11.5 days of gestation. Further analysis of the lethal phenotype, the gene, and its products should elucidate the role of the *Srev-5* locus in embryonic development.

3 Concluding Remarks

Many spontaneous and induced mutations have been identified in the mouse. Complete analysis of the affected genes has often proven difficult because molecular access to these regions is lacking. The use of transposable elements as insertional mutagens has proven immensely useful in dissecting complex processes in several organisms, in particular *D. melanogaster*. In the mouse, retroviruses present in the germline can induce mutations by integrating within or near a gene simultaneously inducing the mutation and providing a tag to molecularly clone the mutated gene. In the case of recessive lethal mutations, the role of the mutated gene in embryonic development can be determined by analyzing the time and circumstances of the lethality. Simultaneously, the affected gene can be molecularly cloned, enabling determination of gene structure, expression, and function. The retrovirally induced mutations identified to date, *Mov-13*, *Mov-34*, *d^v*, *hr*, and *Srev-5*, illustrate the advantages of this approach. For example, the *Mov-13* mutation demonstrated that correct transcription of the $\alpha 1$(I) collagen gene is required at days 12–14 of embryonic development. Further analysis of other retrovirally induced mutations may identify genes required for embryonic viability in the cases of *Mov-34* and *Srev-5*; melanocyte and neurological function as well as embryonic viability in the case of the *dilute* gene; and hair growth, mammary gland function, and thymocyte differentiation in the case of the *hairless* gene.

   Additional retrovirally induced mutations can be produced by either infection of embryos, ES cells, or newborn mice or characterization of existing endogenous proviruses. Introduction of new proviruses into the mouse germline by infection of newborn mice constitutes a novel, technically straightforward method. Infection of either embryos or ES cells are relatively difficult procedures requiring expertise in embryo and cell culture techniques.

As compared to the characterization of existing endogenous proviruses, infection of newborn mice has the additional advantage that viruses can be used that have been altered in vitro, for example, by the addition of a selectable marker to aid in cloning, such as the tRNA gene, *supF*. However, using the newborn infection method, the frequency of acquisition of germline proviruses, although higher than previously reported in other mice, is inadequate for large-scale mutagenesis studies. This difficulty may be solved in two ways. First, the frequency of provirus acquisition may be increased. It is possible that nonecotropic viruses, which use a different cellular receptor for infection, could be·used in combination with virus derived from *Emv-16* and *Emv-17*. It is also possible that mice having a higher susceptibility to germline infection than SWR/J mice may be identified and used. The second solution would be to simplify the methods used to screen mice for newly acquired proviruses and induced mutations. For instance, if the viruses used for infection contained an expressible, dominant, visible phenotypic marker, then mice that acquire the new proviruses could be identified simply by visual examination. This would eliminate the need to screen for the presence of new proviruses by Southern blot analysis of genomic DNA from every mouse produced. In addition, use of screening methods that reveal mutant phenotypes in a single generation would be convenient. One such screen is currently being used to identify mutations that cause sex reversal in the progeny of chimeric mice made from retrovirally infected ES cells (LOVELL-BADGE and ROBERTSON 1990). Alternatively, tester stocks carrying multiple recessive mutations can be used to screen for retrovirally induced mutations in these loci. With the advent of such improvements, in time, retroviral insertional mutagenesis may prove to be as useful a tool in mice as P-element mutagenesis has been in *D.melanogaster*.

Acknowledgements. The authors would like to thank Drs. Peter Donovan, Monica Justice, David Kingsley, Sally Spence, and Marjorie Strobel for helpful discussions. We would also like to thank Drs. Peter Seperack, Sally Spence, and Marjorie Strobel for sharing results that are not yet published. In addition, Debra Gilbert and Deborah Swing contributed greatly to some of the work discussed in this review. We would also like to thank Linda Brubaker and Robin Handley for help in preparing this manuscript.

References

Bautch VL (1986) Genetic background affects integration frequency of ecotropic proviral sequences into the mouse germline. J Virol 60: 693–701

Bradley A, Evans M, Kaufman M, Robertson E (1984) Formation of germline chimaeras from embryo-derived teratocarcinoma cell lines. Nature 309: 255–256

Chattopadhyay SK, Lander MR, Rands E, Lowy DR (1980) Structure of endogenous murine leukemia virus DNA in mouse genomes. Biochemistry 77: 5774–5778

Chen S, Struuck FD, Duran-Reynals ML, Lilly F (1980) Genetic and nongenetic factors in expression of infectious murine leukemia viruses in mice of the DBA/2 × RF cross. Cell 21: 849–855

Copeland NG, Hutchison KW, Jenkins NA (1983) Excision of the DBA ecotropic provirus in dilute coat-color revertants of mice occurs by homologous recombination involving the viral LTRs. Cell 33: 379–387

Evans MJ, Kaufman MH (1981) Establishment in culture of pluripotential cells from mouse embryos. Nature 292: 154–156

Gallagher CH, Path FRC, Canfield PJ, Greenoak GE, Reeve VE (1984) Characterization and histogenesis of tumors in the hairless mouse produced by low-dosage incremental ultraviolet radiation. J Invest Dermatol 83: 169–174

Green MC (1981) Genetic variants and strains of the laboratory mouse. Fischer, Stuttgart

Gridley T, Soriano P, Jaenisch R (1987) Insertional mutagenesis in mice. Trends Genet 3: 162–166

Harbers K, Kuehn M, Delius H, Jaenisch R (1984) Insertion of retrovirus into the first intron of α1(I) collagen gene leads to embryonic lethal mutation in mice. Proc Natl Acad Sci USA 81: 1504–1508

Hutchison KW, Copeland NG, Jenkins NA (1984) Dilute-coat-color locus of mice: nucleotide sequence analysis of the d^{+2J} and d^{+Ha} revertant alleles. Mol Cell Biol 4: 2899–2904

Jaenisch R (1980) Retroviruses and embryogenesis: microinjection of Moloney leukemia virus into midgestation mouse embryos. Cell 19: 181–188

Jaenisch R (1988) Transgenic animals. Science 240: 1468–1474

Jaenisch R, Harbers K, Schnieke A, Löhler J, Chumakov I, Jähner D, Grotkopp D, Hoffmann E (1983) Germline integration of Moloney murine leukemia virus at the Mov13 locus leads to recessive lethal mutation and early embryonic death. Cell 32: 209–216

Jähner D, Jaenisch R (1980) Integration of Moloney leukaemia virus into the germline of mice: correlation between site of integration and virus activation. Nature 287: 456–458

Jenkins NA, Copeland NG (1985) High frequency germline acquisition of ecotropic MLV proviruses in SWR/J–RF/J hybrid mice. Cell 43: 811–819

Jenkins NA, Copeland NG, Taylor BA, Lee BK (1981) Dilute (d) coat colour mutation of DBA/2J mice is associated with the site of integration of an ecotropic MLV genome. Nature 293: 370–374

Jenkins NA, Copeland NG, Taylor BA, Lee BK (1982) Organization, distribution and stability of endogenous ecotropic murine leukemia virus DNA sequences in chromosomes of Mus musculus. J Virol 43: 26–36

Jenkins NA, Strobel MC, Seperack PK, Kingsley DM, Moore KJ, Mercer JA, Russell LB, Copeland NG (1989) A retroviral insertion in dilute (d) locus provides molecular access to this region of mouse chromosome 9. I. Genetic analysis of the d locus. Progr Nucleic Acid Res Mol Biol 36: 207–220

Kuehn MR, Bradley A, Robertson EJ, Evans MJ (1987) A potential animal model for Lesch-Nyhan syndrome through introduction of HPRT mutations into mice. Nature 326: 295–298

Langdon WY, Theodore TS, Buckler CE, Stimpfling JH, Martin MA, Morse HC III (1984) Relationship between a retroviral germ line reintegration and a new mutation at the ashen locus in B10.F mice. Virology 133: 183–190

Lock LF, Keshet E, Gilbert DJ, Jenkins NA, Copeland NG (1988) Studies of the mechanism of spontaneous germline ecotropic provirus acquisition in mice. EMBO J 7: 4169–4177

Lohler J, Timpl R, Jaenisch R (1984) Embryonic lethal mutation in mouse collagen 1 gene causes rupture of blood vessels and is associated with erythropoietic and mesenchymal cell death. Cell 38: 597–607

Lovell-Badge RH, Robertson EJ (1990) XY female mice resulting from a heritable mutation in the primary testis-determining gene, Tdy. Development 109. 635–646

Martin GR (1981) Isolation of a pluripotent cell line from early mouse embryos cultured in medium conditioned by teratocarcinoma stem cells. Proc Natl Acad Sci USA 78:7634–7638

Mayer A, Duran-Reynals ML, Lilly F (1978) Fv regulation of lymphoma development and of thymic, ecotropic and xenotropic MLV expression in mice of the AKR/J × RF/J cross. Cell 15: 429–435

Mayer A, Duran Struuck F, Duran-Reynals ML, Lilly F (1980) Maternally transmitted resistance to lymphoma development in mice of reciprocal crosses of the RF/J and AKR/J strains. Cell 19: 431–436

McCubrey J, Horowitz JM, Risser R (1982) Structure and expression of endogenous ecotropic murine leukemia viruses in RF/J mice. J Exp Med 156: 1461–1474

Meier H, Myers DD, Huebner RJ (1969) Genetic control by the hr-locus of susceptibility and resistance to leukemia. Proc Natl Acad Sci USA 63: 759–766

Morrissey PJ, Parkinson DR, Schwartz RS, Waksal SC (1980) Immunologic abnormalities in HRS/J mice. I. Specific deficit in T lymphocyte helper function in a mutant mouse. J Immunol 125: 1558–1562

Panthier JJ, Condamine H (1987) Expression of ecotropic MuLV in ovaries of SWR/J-RF/J hybrid mice. Ann Inst Pasteur/Virol 138: 409–422

Panthier JJ, Condamine H, Jacob F (1988) Inoculation of newborn SWR/J females with an ecotropic murine leukemia virus can produce transgenic mice. Proc Natl Acad Sci USA 85: 1156–1160

Pincus T, Hartley JW, Rowe WP (1971) A major genetic locus affecting resistance to infection with murine leukemia viruses. I. Tissue culture studies of naturally occurring viruses. J Exp Med 133: 1219–1233

Shih CC, Stoye JP, Coffin JM (1988) Highly preferred targets for retrovirus integration. Cell 53: 531–537

Silvers WK (1979) The coat colors of mice. Springer, New York

Soriano P, Gridley T, Jaenisch R (1987) Retroviruses and insertional mutagenesis in mice: proviral integration at the Mov 34 locus leads to early embryonic death. Genes Dev 1: 366–375

Spence SE, Gilbert DJ, Swing DA, Copeland NG, Jenkins NA (1989) Spontaneous germ line virus infection and retroviral insertional mutagenesis in eighteen transgenic Srev lines of mice. Mol Cell Biol 9: 177–184

Stoye JP, Coffin JM (1987) The four classes of endogenous murine leukemia virus: structural relationships and potential for recombination. J Virol 61: 2659–2669

Stoye JP, Coffin JM (1988) Polymorphism of murine endogenous proviruses revealed by using virus class-specific oligonucleotide probes. J Virol 62: 168-175

Stoye JP, Fenner S, Greenoak GE, Moran C, Coffin JM (1988) Role of endogenous retrovirus as mutagens: the hairless mutation of mice. Cell 54: 383–391

Strobel MC, Seperack PK, Copeland NG, Jenkins NA (1990) Molecular analysis of two dilute locus deletion mutations: the spontaneous dilute-lethal[20J] and the radiation-induced dilute-prenatal-lethal Aa2 alleles. Mol Cell Biol 10/2: 501–509

Taylor BA, Rowe L (1989a) A mouse linkage testing stock possessing multiple copies of the endogenous ecotropic murine leukemia virus genome. Genomics 5: 221–232

Taylor BA, Rowe L (1989b) Mouse sublines bearing novel proviruses. Mouse News Lett 84: 91–92

Varmus HE, Padgett T, Heasley S, Simon G, Bishop JM (1973) Cellular functions are required for the synthesis and integration of avian sarcoma virus-specific DNA. Cell 11: 307–319

Vijaya S, Steffen DL, Robinson HL (1986) Acceptor sites for retroviral integrations map near DNase I-hypersensitive sites in chromatin. J Virol 60: 683–692

Weiss R, Teich N, Varmus H, Coffin J (1984) RNA-tumor viruses, vol 4. Cold Spring Harbor, New York

Insertional Mutagenesis
in Mouse Mammary Tumorigenesis

R. Nusse

1 Mammary Tumors in Mice and Insertional Oncogene Activation by MMTV 44

2 *Wnt-1* 47
2.1 Properties of *Wnt-1* 47
2.2 *Wnt-1* Expression Patterns and Role in Embryogenesis 50
2.3 Developmental Mutations in *Wnt-1* 50
2.4 Assays for *Wnt-1* Activity 51
2.5 *Wnt-1*-Related Genes: *Wnt-2* and *Wnt-3* 52

3 *int-2* 53
3.1 Properties of *int-2* 53
3.2 Expression of *int-2* in Embryos 55
3.3 Assays for *int-2* Activity 55
3.4 *int-2*-Related Genes: *hst* 56

4 *int-3* 57

5 Cooperation Between Different *int* Genes 57

6 Human Tumors 58

7 Speculations and Conclusions 59

References 61

Over the past few years, several surprising discoveries have been made regarding cellular oncogenes which are activated in mouse mammary tumors. These genes provide a particularly good example of the theme of this volume: host cell oncogenes activated by insertion of proviral DNA. They belong especially to this group because most of them have been discovered by virtue of insertional activation, through the application of provirus tagging methods. Quite unexpectedly, what several mouse mammary tumor virus (MMTV) activated oncogenes have in common is that they appear to play key roles in early normal development. In this chapter, the properties of the genes, their products, and their role in embryogenesis will be reviewed (see also Nusse 1988a, b).

Howard Hughes Medical Institute and Department of Developmental Biology, Beckman Center, Stanford University, Stanford, CA 94305, USA

1 Mammary Tumors in Mice
and Insertional Oncogene Activation by MMTV

By inbreeding and selection, several strains of mice have been obtained that are prone to mammary tumorigenesis, even without exposure to carcinogenic chemicals. These mice carry a B-type retrovirus in the milk, MMTV. In some strains of mice, MMTV is present as an endogenous but highly active provirus, capable of inducing tumors even when the milk transmitted virus is removed (reviewed in HILGERS and BENTVELZEN 1979).

Although these in vivo properties of MMTV make it a very useful system for studying the mechanism of mammary carcinogenesis, some peculiarities of the virus have been a major stumbling block in analyzing its genome at the molecular level. MMTV is difficult to grow in culture and there is no good in vitro biological assay for virus activity. Hence, no cloned virus variants are available. Until recently, it was also very difficult to clone MMTV as recombinant DNA, due to a so-called poisonous sequence in its genome (BROOKES et al. 1986). With some tricks the latter difficulties have been solved, leading to the elucidation of the complete nucleotide sequence of MMTV (MOORE et al. 1987) and to some MMTV-based virus vectors (GUENZBURG and SALMONS 1986; SHACKLEFORD and VARMUS 1988; MORRIS et al. 1989). By site-directed mutagenesis of these clones, one should now be able to fully characterize the viral genome. Such manipulations may resolve one of the most pressing questions about the coding potential of MMTV: a gene present in the U3 domain of the viral long terminal repeat (LTR) *orf* (DONEHOWER et al. 1981; KENNEDY et al. 1982; MAJORS and VARMUS 1983), for which no biological activity has been described although a role in transcriptional regulation has been suggested (VAN KLAVEREN and BENTVELZEN 1988).

The tumor-inducing properties of MMTV are intrinsically related to an obligatory step in the retroviral life cycle: insertion of proviral DNA into the host cell DNA (VARMUS 1984). Integration is a mutagenic event for the host cells and one consequence may be the activation of proto-oncogenes whose expression is normally tightly regulated (NUSSE 1986). This is not to say that the whole process of tumorigenesis is caused by a single integration: insertional oncogene activation is an early, if not the first, event (PETERS et al. 1984; MORRIS et al. 1990). Additional damage to the cells, not necessarily initiated by MMTV, may be essential for a fully malignant tumor.

An odd aspect of host gene activation by MMTV, as compared to other retroviruses, is that one rarely finds a gene near MMTV that is known from other sources. Whereas old acquaintances like *myc* are repeatedly encountered at insertional mutations caused by other viruses (see other chapters), there seems to be a specific group of genes associated with MMTV. Those genes, with the exception of *hst* (see below), are usually not activated in other types of tumors and have been discovered by using MMTV DNA as a tag to clone integration domains, followed by chromosome walking and scoring for independent insertions in the same area.

This strategy has now led to the identification of four such common integration domains (Fig. 1, Table 1). The subsequent identification of the oncogene in the cloned domain has been relatively straightforward in the case of *Wnt*-1 (formerly called *int*-1) and *int*-2, the first two genes discovered through this approach. This was facilitated by the multitude of different proviruses found near these genes and by their mode of insertion: proviruses were found at each end of *Wnt*-1 or *int*-2, usually pointing away from the gene itself (Fig. 1). This configuration strongly suggested that the relevant gene would be in between, which proved to be the case: transcripts were found in those tumors with a nearby insertion and the genes were later shown to have growth regulatory properties. Strikingly, all proviral insertions at *Wnt*-1 or *int*-2 leave the protein-encoding domains intact, although quite a few inserts are within the transcription units (VAN OOYEN and NUSSE 1984; NUSSE et al. 1990; DICKSON et al. 1990).

The typical orientation and the large distance of some proviruses from the *Wnt-1* and *int-2* promoters indicate that transcriptional activation of the genes is mediated by enhancers in the MMTV genome. The promoters of both genes have been mapped and are indeed similarly active in tumors as in cells in which the genes are naturally expressed, except for rare cases with proviral insertions within the promoter (NUSSE et al. 1990; DICKSON et al. 1990). For some enhancers it has been documented that they can only act on the first promoter that is encountered in *cis* (WASYLYK et al. 1983), which would explain the typical orientation of MMTV proviruses near *Wnt*-1 and *int*-2 (NUSSE et al. 1984; DICKSON et al. 1984). The opposite orientation would interpose the promoter on the MMTV LTR in between the oncogene promoter and the viral enhancer.

The large distance between MMTV proviruses and the relevant gene has also been a complicating factor in the search for other *int* genes (see GARCIA

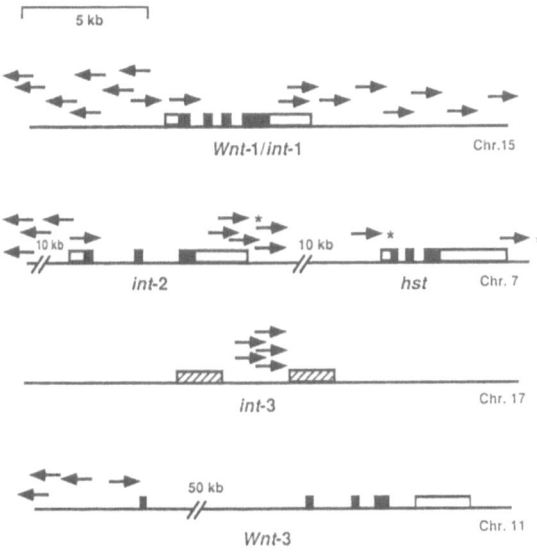

Fig. 1. Organization of the various *int* genes. Four chromosomal areas (chromosome number indicated) are shown where MMTV proviral insertions (*arrows* indicate site and orientation) are found in different mammary tumors. The structure of the genes, where known, is indicated by *boxes; filled areas* correspond to protein-encoding domains. *Stipped boxes* at *int*-3 indicate the position of probes hybridizing to tumor-specific RNA. *Asterisks at arrows* in the *int-2/hst* domain are at MMTV inserts in those tumors where *hst* is expressed

Table 1. Properties of the different *Wnt/int* genes

Gene	Mouse chromosome	Human chromosome	Protein size (kDa)	Related to	Homologue in other species	EC cell expression	Frequency activation in mouse strains
Wnt-1	15	12	44	*Wnt-2, Wnt-3*	Humans (99%) *Xenopus* (68%) *Drosophila* (54%), *C. elegans* Zebrafish	Differentiated P19	80% C3H 70% BR6 30% GR
int-2	7	11	27	FGF	Humans (89%)	Differentiated F9, PCC4, PSA-1	65% BR6 5% C3H 20% GR
int-3	17						40% Czll 8% BR6
Wnt-3	11	17	43	*Wnt-1, Wnt-2*		Differentiated P19	10% GR
hst	7	11	22	FGF	Humans (82%)	Undifferentiated F9, PCC4	10% BR6

et al. 1986). The structure of *int*-3 has been difficult to resolve, in part because there are only a few proviral insertions found near the genes (Fig. 1). At the *int*-2 gene, the situation is even more complex: a common integration area that was first thought to be unrelated to the known *int* genes (SCHUERMANN and MICHALIDES 1987) was later shown to be an extension of the *int*-2 domain (PETERS et al. 1989a). Moreover, it has recently been found that the *int*-2 area contains an additional oncogene, called *hst*, which is transcriptionally activated in some tumors in which MMTV proviruses are found downstream from *int*-2 (Fig. 1; PETERS et al. 1989b). Apart from the interesting biological aspects of these configurations, these findings set instructing caveats for all transposon tagging exercises: tracking down the important gene near an integration area may not be as straightforward as it seems.

Since the tumors where these different genes are activated are morphologi- cally and biologically similar, if not identical, one would expect that the MMTV- activated oncogenes share other properties. They do; as a rule they encode proteins with hallmarks of secretory proteins. In addition, the genes are all expressed in early embryogenesis. With the exception of *int*-3, the genes fall into two groups: the *Wnt* family and the fibroblast growth factor (FGF) related *int*-2 family.

4 *Wnt*-1

2.1 Properties of *Wnt*-1

Wnt-1 is quite frequently activated by MMTV, regardless of the mouse strain of the MMTV variant (NUSSE and VARMUS 1982). The gene is unusually highly conserved in evolution, having a virtually identical human homologue (VAN'T VEER et al. 1984; VAN OOYEN et al. 1985) and easily recognizable amphibian (NOORDERMEER et al. 1989), insect (RIJSEWIJK et al. 1987a) and nematode (KAMB et al. 1989) counterparts (Fig. 2). *Wnt*-1 encodes a protein with many cysteine residues, a signal peptide, and no transmembrane domain (VAN OOYEN and NUSSE 1984; FUNG et al. 1985). These are properties of secretory proteins, and *Wnt*-1 protein, if the gene is overexpressed in various cell lines, indeed enters the secretory pathway (BROWN et al. 1987; PAPKOFF et al. 1987). The protein can be detected in protease-resistant structures, presumably membrane-surrounded secretory organelles, and contains carbohydrate structures at several N-linked glycosylation sites. It is thus generally assumed that the *Wnt*-1 protein is secreted from cells, but extracellular forms of the protein have nevertheless been difficult to detect (PAPKOFF et al. 1987; SCHUURING et al. 1990). In addition, most of the intracellular *Wnt*-1 protein is incompletely glycosylated, as it remains sensitive to endoglycosidase H, and has probably not been passaged through the Golgi apparatus (PAPKOFF et al. 1987). More recently, it has been shown that *Wnt*-1

```
Wingless   MDISYIFVIC LMALCSGGSS LSQVEGKQKS GRGRGSMWWG IAKVGEPNNI
    Wnt1       MGLWAL LPSWVSTTLL LALTALPAAL AASSSGRWWG IVNIASSTNL
    Wnt2        MNVPLGGIW LWLPLLLTWL TPEVNSSWWY MRATGGSSR.
    Wnt3         MEPHLLGL LLGLLLSGTR VLAGYPIWWS LALGQQYTSL
                            *                   **

Wingless   .T...PIMYM DPAIHSTLRR KQRRLVRDNP GVLGALVKGA NLAISECQHQ
    Wnt1   LTDSKSLQLV LEPSLQLLSR KQRRLIRQNP GILHSVSGGL QSAVRECKWQ
    Wnt2   .........V MCDNVPGLVS RQRQLCHRHP DVMRAIGLGV AEWTAECQHQ
    Wnt3   ASQP.....L LCGSIPGLVP KQLRFCRNYI EIMPSVAEGV KLGIQECQHQ
                    *              *          *       **

Wingless   FRNRRWNCSTRNFSRGKNLFGKI VD.RGCRETS FIYAITSAA VTHSIARACSE
    Wnt1   FRNRRWNCPT APGP..HLFGK.IVN.RGCRETA FIFAITSAG VTHSVARSCSE
    Wnt2   FRQHRWWNCNT LDRD.HSLFGR..VLLRSSRESA FVYAISSAG VVFAITRACSQ
    Wnt3   FRGRRWNCTT IDDS.LAIFGR..VLDKATRESA FVHAIASAG VAFAVTRSCAE
           ** ****  *       **   *     **   * ** **   *    * *

Wingless   GTIESCTCDYSHQSRSPQ ANHQAGSVAG VRDWEWGGCS DNIGFGFKFS
    Wnt1   GSIESCTCDY RRRGP..... .......GGPDWHWGGCS DNIDFGRLFG
    Wnt2   GELKSCSCDP KKKGSAK... ....DS..KGFFDWGGCS DNIDYGIKFA
    Wnt3   GTSTICGCDS HHHKGP..... ......P GEGWKWGGCS EDADFGVLVS
           *    * **              ***** *

Wingless   REFVDTGER. GRNLREKMNL HNNEAGRAHV QAEMRQECKC HGMSGSCTVK
    Wnt1   REFVDSGEK. GRDLRFLMNL HNNEAGRTTV FSEMRQECKC HGMSGSCTVR
    Wnt2   RAFVDAKERK GKDARALMNL HNNRAGRKAV KRFLKQECKC HGVSGSCTLR
    Wnt3   REFADAREN. RPDARSAMNK HNNEAGRTTI LDHMHLKCKC HGLSGSCEVK
           * * *  *         *  **  *** ***      *** ** ****

Wingless   TCWMRLANFR VIGDNLKARF DGATRVQVTN ..SLRA.TNA LAPVSPNAAG
    Wnt1   TCWMRLPTLR AVGDVLRDRF DGASRVLYGN RGSNRASRAE ..........
    Wnt2   TCWLAMADFR KTGEYLWRKY NGAIQVVMNQ DGTG...... ..........
    Wnt3   TCWWAQPDFR AIGDFLKDKY DSASEMVVEK HRESRG.WVE TLR.......
           ***       *     * *       *

Wingless   SNSVGSNGLI IPQSGLVYGE EEERMLNDHM PDILLENSHP ISKIHHPNMP
    Wnt1   .......... .......... .......... .......... ..........
    Wnt2   .......... .......... .......... .......... ..........
    Wnt3   .......... .......... .......... .......... ..........

Wingless   SPNSLPQAGQ RGGRNGRRQG RKHNRYHFQL NPHNPEHKPP GSKDLVYLEP
    Wnt1   .......... .......... ......LLRL EPEDPAHKPP SPHDLVYFEK
    Wnt2   .......... .......... .........F TVANKRFKKP TKNDLVYFEN
    Wnt3   .......... .......... .......... .AKYALFKPP TERDLVYYEN
                                              * *     **** *

Wingless   SPSFCEKNLR QGILGTHGRQ CNETSLGVDG CGLMCCGRGY RRDEVVVVER
    Wnt1   SPNFCTYSGR LGTAGTAGRA CNSSSPALDG CELLCCGRGH RTRTQRVTER
    Wnt2   SPDYCIRDRE AGSLGTAGRV CNLTSRGMDS CEVMCCGRGY DTSHVTRMTK
    Wnt3   SPNFCEPNPE TGSFGTRDRT CNVTSHGIDG CDLLCCGRGH NTRTEKRKEK
           **  *      * **  *  **  *  *  *    *****

Wingless   CACTFHWCCE VKCKLCRTKK VIYTCL
    Wnt1   CNCTFHWCCH VSCRNCTHTR VLHECL
    Wnt2   CECKFHWCCA VRCQDCLEAL DVHTCKAPKS ADWATPT
    Wnt3   CHCVFHWCCY VSCQECIRIY DVHTCK
           * * *****  * *  *           *
```

overproduction leads to small amounts of extracellular protein which has undergone more extensive glycosylations (PAPKOFF 1989) and may bind to the cell surface (PAPKOFF and SCHRYVER 1990) or to the extracellular matrix (BRADLEY and BROWN 1990), possibly explaining its absence in free form in tissue culture medium. The cell surface-associated Wnt-1 protein can be released by treating cells with suramin (PAPKOFF and SCHRYVER 1990).

One of the problems in detecting the Wnt-1 protein is the low affinity of the available antibodies. This can partially be circumvented by inserting a strongly antigenic epitope into the protein, hoping that such an insertion will not affect the behavior of the protein. By inserting an epitope of the myc protein, to which particularly good monoclonal antibodies have been generated, the Wnt-1 protein could be visualized in overproducing cells, but again only in the endoplasmic reticulum and not outside the cells (McMAHON and MOON 1989a).

A similar but natural solution to the problem of weak antigenicity of Wnt-1 has been found in Drosophila, Where Wnt-1 is the same gene as the segment polarity gene wingless (RIJSEWIJK et al. 1987a; BAKER 1987). The mouse and Drosophila Wnt-1 proteins are identical in 54% of their amino acids, but the Drosophila gene has an insert encoding an additional 85 amino acids. Antibodies raised to the D int-1-wingless protein are able to detect the protein in intact embryos and in individual cells, and it is probably the insert that has provided the highly antigenic determinant. Secretion of the wingless protein has been observed by electron microscopy; the protein is seen in small membrane-bound vesicles, in multivesicular bodies, and in the intercellular space (VAN DEN HEUVEL et al. 1989). Traveling of the protein is suggested by an experiment employing double staining of wild-type embryos with both wingless and engrailed antibodies. The engrailed protein is located in the nucleus of cells immediately posterior to the wingless cells. The multivesicular bodies containing the wingless protein are occasionally found in engrailed positive cells, suggesting that the wingless protein behaves as a paracrine signal (VAN DEN HEUVEL et al. 1989).

All in all, it seems that Wnt-1 protein is indeed secreted like any other secretory molecule. When the gene is overexpressed, however, proper folding of the protein may not occur, resulting in retention as an aggregate in the endoplasmic reticulum (ER). This aggregation has frustrated many attempts to purify biologically active Wnt-1 protein from mammalian or insect cells, although high yields of protein have been obtained (H.E. VARMUS personal communication).

Experimental evidence for a Wnt-1 receptor has not been obtained, but, given the secretion of the protein, we can expect that the Wnt-1 receptor will behave like other growth factor receptors: binding its ligand on the cell surface followed by internalization.

◄───

Fig. 2. Amino acid sequences of the mouse Wnt-1, Wnt-2, and Wnt-3 and the wingless gene. Positions of identical amino acids are indicated by an asterisk. Positions of common introns shown by arrows pointing down, position of unique Drosophila intron by arrow pointing up. (From VAN OOYEN and NUSSE, 1984; RIJSEWIJK et al. 1987a, b; McMAHON and McMAHON 1989; and ROELINK et al. 1990)

2.2 *Wnt*-1 Expression Patterns and Role in Embryogenesis

In adult mice, *Wnt*-1 is not expressed in any tissue, except for the mature testis (JAKOBOVITZ et al. 1986; SHACKLEFORD and VARMUS 1987). Expression is localized in postmeiotic round spermatids, but the biological role of this expression is unclear. In mouse embryos, *Wnt*-1 is expressed between days 8 and 14, specifically in the developing nervous system (WILKINSON et al. 1987; DAVIS and JOYNER 1988). Areas in the neural plate, the anterior head folds, the folding neural tube, and the spinal cord show a highly localized expression of *Wnt*-1. During folding of the neural tube, expression moves from anterior to posterior (WILKINSON et al. 1987). Also, in *Xenopus* embryos, *Wnt*-1 expression is associated with neural development (NOORDERMEER et al. 1989). It has been suggested that the gene is involved in positional signaling in the developing neural system, providing information specifying the midline axis (MCMAHON and MOON 1989a). In line with this view are the dramatic effects on developing *Xenopus* embryos when *Wnt*-1 RNA is injected into fertilized eggs: a bifurcating anterior neural tube and expansion of the posterior neural plate. In addition, the underlying axial mesoderm is duplicated, suggesting dual formation of the whole embryonic axis, perhaps as a result of interference with the so-called organizer (MCMAHON and MOON 1989b).

Most tested embryonal carcinoma cell lines do not express *Wnt*-1, except for P19 cells (JONES- VILLENEUVE et al. 1983) induced by retinoic acid to differentiate along the neural pathway (SCHUURING et al. 1989; ST-ARNAUD et al. 1989).

2.3 Developmental Mutations in *Wnt*-1

Wnt-1 was one of the first genes in which deliberate mutations were made in the mouse germ line. By homologous recombination in mouse embryonic stem cells, a null mutation in the gene was generated (K. THOMAS and M. CAPPECHI; A. MCMAHON, personal communications). Subsequent transfer of the manipulated stem cells to blastocysts has led to germ line transmission of the inactive allele. Homozygous offspring animals have severe ataxia and are barely viable. Embryos at day 17 have an underdeveloped midbrain and cerebellum, but apparently a normal forebrain and spinal cord.

When the *Drosophila* homologue of *Wnt*-1 was cloned, it turned out to be identical to the segment polarity gene *wingless* (BAKER 1987; RIJSEWIJK et al. 1987a; UZVOLGY et al. 1988). From clonal analysis of *wingless,* cells, it appeared that the phenotype is nonautonomous in mosaics, which suggests that the protein has a role in intercellular signalling (MORATA and LAWRENCE 1977). Various alleles of *wingless* are known. A viable allele, which gave the gene its name, leads to a homeotic transformation of the wing into a notum. Embryonic lethal alleles cause more severe effects, such as the absence of segment boundaries and an inverted polarity of the posterior zone of each segment (BAKER 1987). Expression of *wingless* has been localized both with in situ RNA hybridization and with immunostaining. The gene is expressed in regions which

Table 2. Segment polarity genes; phenotypes and structure of products[a]

Phenotype	Gene	Protein structure[b]
Naked-like	naked	?
	zeste-white-3	?
Patched-like	patched	Multiple transmembrane protein
	costal-2	?
Wingless-like	wingless	Secreted factor
	armadillo	Cytoskeletal protein
	dishevelled	?
	porcupine	?
	fused	?
	gooseberry	Nuclear (homeobox) protein
	hedgehog	?
	Cell	?
	Cubitus-interruptus[D]	?
Engrailed-like	engrailed	Nuclear (homeobox) protein

[a] Based on Perrimon and Smouse (1989)
[b] Structure mostly deduced from the sequence of the gene

are aberrantly developed in *wingless* mutants: different regions of the head, the hindgut/anal region and in 15 stripes in the trunk region. After gastrulation, expression is seen in each segment in a few cells just anterior to the parasegment boundary (BAKER 1987; VAN DEN HEUVEL et al. 1989).

The genetics of *Drosophila* segmentation may offer an alternative approach to find a *Wnt*-1 receptor: the phenotype of a *wingless* receptor mutant may be similar to *wingless* itself, and at least eight genes with such a phenotype have been characterized (Table 2). Some of those mutants behave in a cell-autonomous way, which would also be expected of a receptor gene (RIGGLEMAN et al. 1989). As many of these genes are now being cloned (reviewed in INGHAM 1988), more information will soon become available. Further genetic and biochemical analysis of the role of *wingless* in segmentation and outgrowth of imaginal disks will undoubtedly help to elucidate the signal transduction pathway in which the gene is involved, with obvious ramifications for the mechanism of action of *Wnt*-1 in the mouse.

Several circuits of such interactions have been proposed including expression of the nuclear protein *engrailed* in target cells (DiNARDO et al. 1988; MARTINEZ ARIAS et al. 1988), transcriptional repression of *wingless* by the products of the pair-rule genes even-skipped and *fushi-tarazu* (INGHAM et al. 1988), and inhibition of *wingless* protein action by expression of the membrane protein patched (NAKANO et al. 1989).

2.4 Assays for *Wnt*-1 Activity

There are now several biological assays based on *Wnt*-1 activity. The most dramatic effects of the gene are seen in vivo in intact animals. The effects of ectopic expression of *Wnt*-1 RNA in *Xenopus* embryos have already been mentioned. In the mouse a transgene consisting of *Wnt*-1 with an MMTV LTR inserted upstream and in the opposite orientation to the gene, causes massive proliferation of mammary epithelial cells, resulting in hyperplastic glands (TSUKAMOTO et al. 1988). Those mice are not able to nurse their offspring. After a period of latency, mammary tumors emerge in a stochastic pattern, presumably by somatic mutations. The hyperplasia in *Wnt*-1 transgenic mice also occurs in males, which is a unique aspect of *Wnt*-1 male mice carrying other oncogenes linked to the MMTV LTR are usually normal.

Transgenic flies with the *wingless* gene under a heat shock promoter display interesting phenotypes during embryogenesis. The most extreme phenotype is a naked cuticle, which is, not unexpectedly, the opposite of the *wingless* phenotype (J. NOORDERMEER, P. LAWRENCE and R. NUSSE, unpublished work).

The in vitro effects of *Wnt*-1 are less spectacular. By retroviral or direct gene transfer, two cell lines of mammary origin can be morphologically transformed. These cell lines, C57MG and RAC, normally have a cuboidal phenotype, grow to a low density, and are not tumorigenic. Upon uptake and overexpression of *Wnt*-1 constructs, the cells become elongated, from foci, and in the case of RAC cells grow as tumors in suitable recipient mice (BROWN et al. 1986; RIJSEWIIK et al. 1987b). These effects are not seen when fibroblastic lines are used and hint to a cell type specificity of *Wnt*-1 transforming ability. A problem with these assays is that the transformed cells are quite unlike mammary tumor cells with activated *Wnt*-1; they lack most markers of mammary cells and the tumors growing out of the transfected RAC cells are pathologically different from primary mouse mammary tumors (RIJSEWIJK et al. 1987b; SONNENBERG et al. 1987; SCHUURING et al. 1990). The full spectrum of *Wnt*-1 activity may thus not be completely represented by these in vitro transformed cells.

2.5 *Wnt*-1-Related Genes; *Wnt*-2 and *Wnt*-3

Wnt-1 is part of a growing family of genes, which may comprise ten members in the mouse (Figs. 1, 2) (A. MCMAHON, personal communication; H. ROELINK, unpublished work). One of these genes, *Wnt*-3, has been identified by provirus tagging; others have been found through homology with *Wnt*-1. *Wnt*-3 is rather infrequently activated by MMTV (ROELINK et al. 1990). The size of the transcriptional unit, which appears to cover more than 50 kb, has complicated its structural analysis. The gene has been mapped on chromosome 11 in the mouse as part of a cluster of genes which are conserved on human chromosome 17 as a linkage group (BUCHBERG et al. 1989; ROELINK et al. 1990) *Wnt*-3 is normally expressed in mouse embryos as a 3.8 kb mRNA, probably in the developing brain, and in differentiating P19 embryonal carcinoma cells. The amino acid sequence of *Wnt*-3 is 47% identical to *Wnt*-1, with most cysteines conserved (ROELINK et al. 1990; Fig. 2).

The *Wnt*-2 gene was encountered by WRAINWRIGHT et al. (1988), who were in search of a gene predisposing to cystic fibrosis (CF), on human chromosome 7 by a chromosome walk. *Wnt*-2 is not the CF gene, but shows, 38% identity with *Wnt*-1 with most cysteines conserved, hence the name *Wnt*-1 related protein.

A mouse homologue of *Wnt*-2 has been isolated and mapped on chromosome 6 (CHAN et al. 1989; MCMAHON and MCMAHON 1989). Expression of mouse *Wnt*-2 is developmentally regulated, with early expression around day 9 in the pericardium of the heart, the umbilical vein, and the associated allantoic mesoderm. In adults there is expression of *Wnt*-2 in the placenta, the lungs, and the heart and also in human tissue. A role for *Wnt*-2 in the development of fetal allantoic communication has been suggested. Interestingly, *Wnt*-2 expressed before day 9 appears to be differentially spliced, since mRNA is found for a protein lacking a signal peptide, whereas later in development a protein with such a leader can be made (MCMAHON and MCMAHON 1989).

In view of the homology with *Wnt*-1, we have examined whether *Wnt*-2 can also behave as an oncogene in mouse mammary tumors. No tumors with MMTV insertions near the gene were found, but several tumors appeared to have amplified copies of *Wnt*-2. In these tumors, which were negative for *Wnt*-1 activation, considerable overexpression of *Wnt*-2 was detected (unpublished observations).

3 *int*-2

3.1 Properties of *int*-2

The frequency of proviral activation of *int*-2 is similar to *Wnt*-1, but, for reasons unknown, some strains of mice, in particular the BR6 strain, have a higher incidence of *int*-2 activation than others, e.g., the C3H strain (PETERS et al. 1983, 1986). *int*-2 is conserved in evolution, with homologues in humans (BROOKES et al. 1989a) and in XENOPUS (C. DICKSON, personal communication). Transcription of *int*-2 is complex; there are at least six mRNAs found in tumors or in embryonal carcinoma cell lines, resulting from differential use of promoters and polyadenylation signals (SMITH et al. 1988; MANSOUR and MARTIN 1988). In some tumors, the activating provirus structurally altered the *int*-2 initiation sites (DICKSON et al. 1990). Nevertheless, all of these transcripts, whether from mammary tumors or from embryonal carcinoma cell lines, can give rise to the synthesis of the same protein, which shares substantial homology with the FGF family of growth factors (MOORE et al. 1986; DICKSON and PETERS 1987; SMITH et al. 1988). Like some, but not all, members of this family (Fig. 3, see ROGELJ et al. 1988), the *int*-2 protein has a signal peptide and no transmembrane domain, implicating a secreted product; however direct evidence that the *int*-2 protein is secreted is lacking. By immunoblotting experiments, *int*-2 proteins have been identified in COS cells overexpressing a copy of the gene which has been engineered to

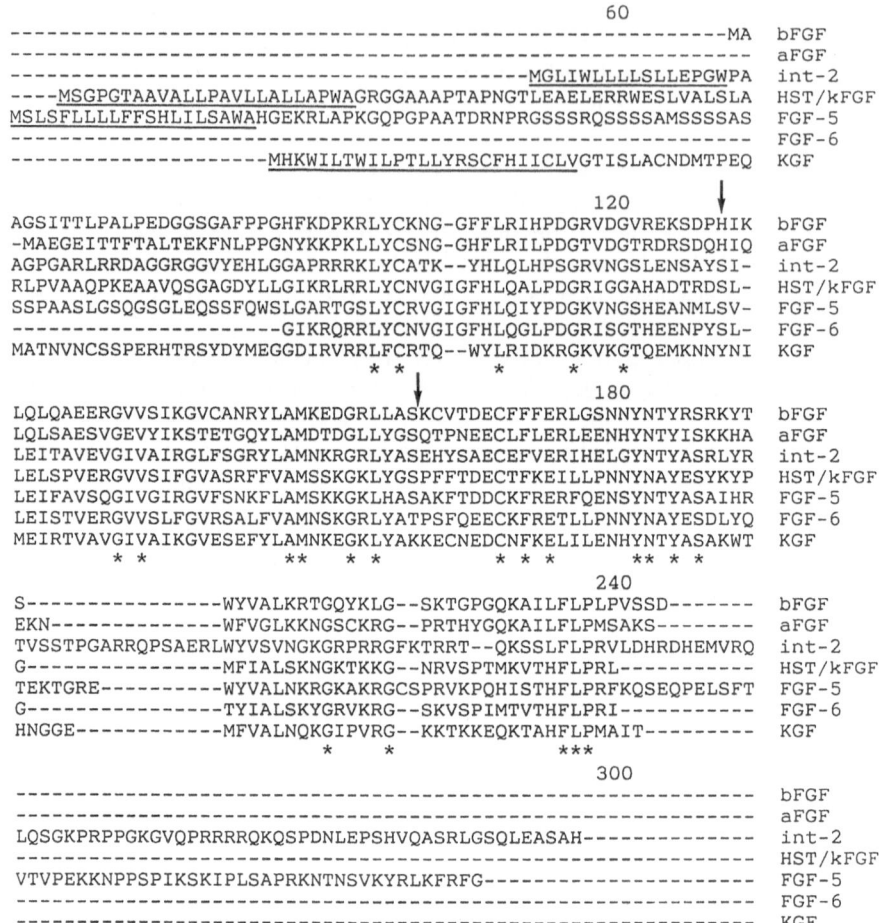

Fig. 3. Amino acid sequences of all human FGF-like genes. Hydrophobic signal sequences are *underlined*, identical amino acids indicated by an *asterisk*, and positions of introns which are common for all members indicated by an *arrow*. (From Jaye et al. 1986; Abraham et al. 1986; Delli-Bovi et al. 1987; Taira et al. 1987; Brookes et al. 1989a, b; Zhan et al. 1988; Marics et al. 1989, Finch et al. 1989)

optimalize protein synthesis (Dixon et al. 1990). Four species are found, ranging from 27.5 to 31.5 kDa and differing in extent of glycosylation, with additional variance due to incomplete removal of the signal peptide. The protein enters microsomes supplied to in vitro translation assays. In immunofluorescence experiments, the protein can be seen in the ER and in the Golgi. Since the available antibodies failed in immunoprecipitation assays and may have a low affinity, a possibly secreted form of the *int*-2 protein may have escaped detection altogether. Alternatively, the *int*-2 proteins may not be destined to travel far from the producing cell and bind tightly to cell surface structures.

Surprisingly, *int*-2 protein synthesis can utilize a CUG initiation codon that is upstream and in frame with the more conventional AUG initiation site. Transfection of *int*-2 constructs leading to either the normal or the amino-extended form of the protein shows that the latter form enters the nucleus, whereas the AUG-initiated product is found in the secretory pathway (ACLAND et al. 1990). The nuclear uptake of the extended, CUG initiated protein is probably mediated through a nuclear location signal that is present in both forms of the protein and perhaps unmasked by the amino-terminal extension.

3.2 Expression of *int*-2 in Embryos

From its expression pattern in mouse embryos, essential roles for the *int*-2 gene product in various phases of development are apparent. By in situ RNA hybridizations, expression has been detected as early as day 7.5 of the developing mouse embryo (WILKINSON et al. 1988). Most intriguing is the expression seen in presumptive mesodermal cells migrating through the primitive streak during gastrulation and in migrating parietal endodermal cells. Expression has also been detected at specific sites in the developing neuroepithelium. Later in development, expression can be seen in the cerebellum and in the developing retina, in the inner ear, and in the developing teeth (WILKINSON et al. 1989). There is no correlation with sites of high cell proliferation; expression of *int*-2 is associated with cell migration or induction events. These findings, made at the same time when it was discovered that other FGF-like proteins have mesoderm induction properties in amphibian embryos (SLACK et al. 1987; KIMELMAN and KIRSCHNER 1987), strongly indicate that the *int*-2 protein may have a decisive role in embryonic development, as a local cell-cell interaction molecule. In *int*-2 protein made in vitro is indeed also able to induce mesoderm in frog embryos, albeit less efficiently than basic FGF or *hst* (PATERNO et al. 1989).

Many embryonal carcinoma cell lines, most notably F9 and PCC4 cells, express *int*-2 upon differentiation, in particular in the endodermal lineage (JAKOBOVITS et al. 1986). Various *int*-2 mRNAs are seen, differing in sites of initiation and polyadenylation (SMITH et al. 1988).

3.3 Assays for *int*-2 Activity

Transfections with activated *int*-2 constructs in NIH 3T3 cells lead to small but reproducible changes in cell morphology and growth rate (C. DICKSON and G. PETERS, personal, communication). Interestingly, there is also a measurable effect of in vitro made *int*-2 protein. When added to growth-arrested fibroblastic or C57MG mammary epithelial cells, the *int*-2 protein is weakly mitogenic (DIXON et al. 1990). Such assays will be very useful in isolating the *int*-2 protein in a biologically active form, with mesoderm induction in *Xenopus* embryos as additional criterion. *int*-2 seems to have a wider range of target cells than *Wnt*-1, which can only transform mammary epithelial cells (BROWN et al. 1986; RIJSEWIJK et al. 1986).

Transgenic mice have been made which carry the *int*-2 gene driven by the MMTV promoter/enhancer (MULLER et al. 1990). Expression of *int*-2 in female transgenic mice results in pronounced mammary gland hyperplasia. Interestingly, expression of the *trans*-gene in the prostate gland of male carriers results in a benign epithelial hyperplasia, indicating that the *int*-2 product can act as a potent growth factor in several epithelial tissues.

A receptor specific for the *int*-2 protein has not been identified as yet, but with the growth stimulatory assay for in vitro made protein in hand, it should be feasible to identify such a molecule. The chicken receptor for bFGF has recently been cloned; the protein appears to be a transmembrane molecule with extracellular immunoglobulin domains and an intracellular tyrosine kinase domain (LEE et al. 1989). It is of obvious interest to determine to what exent this receptor can bind the FGF-related proteins, including *int*-2.

3.4 *int*-2 Related Genes: *hst*

The *hst* gene is the only example of an oncogene activated by MMTV that was known from previous work. By transfecting genomic DNA from either a Kaposi sarcoma (DELLI-BOVI et al. 1987) or a stomach carcinoma (M.C. YOSHIDA et al. 1988; T.K. YOSHIDA et al. 1987; TAIRA et al. 1987) a strongly transforming oncogene was found with homology to the FGF family. The *hst* protein is equipped with a signal peptide and is glycosylated and transported efficiently to the outside of cells, where it can be stabilized by added heparin (DELLI-BOVI et al. 1988). Conditioned medium from transfected cells contains a growth stimulatory activity, presumably the *hst* protein itself, and it has even been possible to produce biologically active *hst* protein in bacteria (DELLI-BOVI et al. 1987).

The possible role of *hst* in mouse mammary tumors was first suggested by its homology to FGFs, and thereby to *int*-2, and then by its genomic location. Both in the mouse and in the human genome, *hst* and *int*-2 are closely linked (HUEBNER et al. 1988). In the mouse, the distance from *int*-2 to *hst* is only 17 kb (PETERS et al. 1989b; BROOKES et al. 1989b). Of the MMTV proviral insertions mapped near *int*-2, several were rather far from the gene itself, and in some of these no *int*-2 transcription had been detected. It now appears that these tumors express the *hst* gene, activated by MMTV insertions at either side (PETERS et al. 1989b; Fig. 1). Moreover, in some of these tumors, both *int*-2 and *hst* are expressed, which is the more intriguing since these two genes are related to each other, each being a member of the FGF family, and may act in the same molecular pathway.

Surprisingly, mouse mammary tumors are thus far the only tumors where *hst* is implicated in tumorigenesis. The normal human *hst* gene is a potently transforming oncogene in NIH 3T3 cells without manipulation, but there is no evidence that the gene had any causal relationship with the tumors it was originally isolated from (SAKAMOTO et al. 1988).

4 *int*-3

The structure of the *int*-3 gene is not completely analyzed yet, nor is its normal pattern of expression known (GALLAHAN and CALLAHAN 1987; GALLAHAN et al. 1987). The most intriguing properties of the gene are the strong association with tumors in the Czech-2 strain of mice (Table 1) and the configuration of inserted proviruses with respect to the gene (Fig. 1). All ten proviral insertions found at *int*-3 thus far map apparently in the middle of a gene, disrupting the transcriptional unit in such a way that the upstream transcription is terminated in the left LTR of the proviruses and the downstream part of the gene is activated by promoter insertion from the right LTR (R. CALLAHAN, personal communication). It is not known which part of the gene is essential for tumorigenesis. The configuration is reminiscent of the avian leukemia virus (ALV) insertions near the chicken epidermal growth factor (EGF) receptor gene (or c-*erb*-B) that generate a truncated receptor (see H.-J. KUNG, C. BOERKOEL and T. CARTER, this volume). Partial sequence information for *int*-3 did not reveal any homology with known genes.

5 Cooperation Between Different *Int* Genes

Most provirus tagging exercises are started in a tumor with a single acquired provirus, since there one has a fair chance of cloning a relevant integration site. The majority of tumors, however, have multiple insertions. Are the additional proviruses innocuous, inserted into the tumor clone before the oncogenic mutation occurred, or have they actively played a role in tumor formation? With the identification of more *int* genes, many examples have now been found in which more than one gene is activated in a single tumor. In a set of 35 BR6 tumors, PETERS et al. (1986) found 50% of the samples to have *Wnt*-1 plus *int*-2 rearranged, most likely in a clonal fashion. Another example are those tumors in which *int*-2 and *hst* are both activated, albeit by a single proviral insertion. In our laboratory we have seen tumors in which *Wnt*-3 is activated that also have amplified copies of the *Wnt*-2 gene, with correspondingly high levels of *Wnt*-2 RNA. Some early arising tumors in the GR strain, that are still completely dependent on pregnancy hormones for growth are intriguing. These tumors may have multiple *int* genes activated, but now in different clones within the same tumor. The clonal balance of these subpopulations is maintained when the tumor is transplanted, suggesting mutual interactions of the clones (GRAY et al. 1986; MESTER et al. 1987). Such interactions may well be mediated directly by the *int* proteins acting as paracrine growth factors.

These observations suggest that different genes can cooperate in tumorigenesis. Indeed, a single gene activation is not sufficient for a fully malignant tumor; transgenic mice carrying a germline copy of *Wnt*-1 which is transcriptionally active in many if not all mammary cells, develop tumors in a stochastic way

(TSUKAMOTO et al. 1988). This indicates that additional mutations not caused by MMTV, which is absent from these mice, cause clonal outgrowth of the tumors. The nature of these events is unknown since transcriptional activity of *int*-2 has not been detected. Unexpectedly, the tumors in transgenic animals emerge after approximately the same period of latency as tumors in which *Wnt*-1 is somatically activated by infectious MMTV. Possibly the virus has an accessory function in tumorigenesis by providing a gene product that by itself is not sufficient to cause a tumor (otherwise tumors would not be clonal), but that can cooperate with the oncogene product.

6 Human Tumors

One of the goals of oncogene research in mammary cancer is to determine the possible involvement of oncogenes in human breast cancer, especially since no known oncogene has been implicated in the majority of human mammary tumors. Human *Wnt*-1 and *int*-2 homologues have been cloned and sequenced and are highly similar to their mouse counterparts (Table 1). A systematic screen for rearrangements or overexpression of these homologues has yielded disappointing results.

Wnt-1 is not amplified, translocated, or even expressed in human breast cancer (M VAN DE VIJVER and R. NUSSE, unpublished work). Matters are more complicated for *int*-2, which is amplified in quite a few (30%) human mammary cancers. (ZHOU et al. 1988; VARLEY et al. 1988; FANTL et al. 1989). The degree of amplification is low and is not paralleled by expression of the gene (FANTL et al. 1989). In most of the tumors with *int*-2 amplification, a whole cluster of adjacent genes, including the oncogenes *hst* and *bcl*-1, is coamplified (ADELAIDE et al. 1988; HUEBNER et al. 1988; LIDEREAU et al. 1988; TSUDA et al. 1989; M.C. YOSHIDA et al. 1988; TSUTSUMI et al. 1988). Such large regions of amplified DNA are generally seen when only one gene is driving the amplification by being overexpressed and conferring a selective growth advantage. In the case of the *Int*-2 cluster, however, none of the known genes have been shown to be expressed, and it is thought that some as yet unidentified mammary oncogene is hidden in this cluster.

It is perhaps not to be expected that the *int/Wnt* genes are amplified in human mammary tumors. For a gene to become overexpressed by amplification, a low level of normal expression is required and *Wnt*-1 and *int*-2 are not detectably expressed in normal mammary gland tissue. Therefore, if there is any role for the *int/Wnt* genes in human tumors, it is perhaps more likely to be found in tumors of embryonic origin, for example the neuroectodermal lineage. In this light, the observation of ARHEDEN et al. (1989) that *Wnt*-1 is amplified in human retinoblastomas is provocative, although expression of the gene has not been examined.

7 Speculations and Conclusions

The *int/Wnt* genes are one of the best examples of a link between cancer and the control of normal early development. Cancer would then be the consequence of erroneous expression of genes whose normal function is confined to decision-making during embryogenesis. The evidence that these genes normally have such a role is compelling: restricted expression, induction capacities in frog embryos, and identity with developmental genes. In embryogenesis the *int* genes may function in pattern formation by encoding secreted proteins with a short range of action, influencing the fate of surrounding cells or tissues.

By summarizing what is known of the *Wnt/int* genes, one becomes intrigued as to whether there is an underlying mechanism that determines the specificity of these developmental genes for mammary cancer induced by MMTV. Why are *myc* or c-*erb* B-2 never activated by MMTV insertion, whereas these and many other oncogenes are capable of inducing mouse mammary tumors if present as transgenes (LEDER et al. 1986; MULLER et al. 1988; BOUCHARD et al. 1989)? One explanation could be that the MMTV enhancer is to some extent specific for certain promoters and that the virus-activated oncogenes, all expressed in embryogenesis, would share these promoter elements. The patterns of expression of the different genes, however, do not indicate that they have such common regulatory elements. For example, *int*-2 and *hst* expression in F9 embryonal carcinoma cells (Table 1) are opposite to each other when the cells are differentiated.

Critical to the question of a common mechanism is the level of expression in mammary tissue of the *Wnt*/int genes, which is below detection before proviral activation and yet not higher than 20 copies of RNA per cell after a proviral hit (FUNG et al. 1985). Other proto-oncogenes are usually expressed to at least such levels in normal cells; they may only become transforming if this level is elevated to an extent that cannot be reached by MMTV enhancement. Apparently, the mouse mammary gland is extremely sensitive to the low but inadvertent expression of an *int/Wnt* gene. This would imply that mammary cells have receptors for the *int/Wnt* genes. In turn, the presence of these receptors would suggest that a member of the *int/Wnt* family is a normal growth factor for the mammary gland. In principle, these factors could be identical to *Wnt*-1 or *int*-2, the oncogenes themselves, but available evidence suggests that they are not at all expressed during mammary gland growth. More likely therefore, the normal growth factors are the products of related genes, able to bind to the same receptor as the oncogene products. Such a factor is then normally regulated during the growth and development of the mammary gland and downregulated after growth. The oncogenic activation of *Wnt*-1 or *int*-2 would thus take over the action of these factors. Since both the *Wnt* and the FGF family appears to be large, consisting of some ten members, there are many possible candidates that could act as natural mammary growth factors.

In addition, developmental genes can be potent mammary oncogenes since the tumors arise in a developing organ: in mammary glands cycling through multiple rounds of pregnancy. In this aspect, the mammary gland is unique among the adult organs, which may explain the unique association of abnormal mammary growth with ectopic activation of developmental genes. The major mitogenic stimulus in tumorigenesis comes then from pregnancy hormones, and aberrant growth may result from activation of genes controling normal growth at an abnormal point in life.

Another mechanism suggested by the specific biology of the mammary gland is delay of cell death after lactation, when the gland regresses. During a subsequent round of gland development, cells with a prolonged lifespan would still be present, have a growth advantage, and be prone to secondary tumorigenic events. In this respect, it is interesting to note that basic FGF can delay cell senescence (GOSPODAROWICZ et al. 1987) and that lack of *wingless* function has been suggested to lead to cell death (PERRIMON and MAHOWALD 1987). The hypothesis is also fostered by the observation that *Wnt*-1 as a transgene leads also in male mice to aberrant mammary gland development (TSUKAMOTO et al. 1988). Normally, the male mammary gland develops initially like that of the female, but the gland regresses under the influence of male hormones acting on the surrounding mesenchyme, which in turn secretes substances destroying the gland (KRATOCHWIL 1985). The *Wnt*-1 transgene apparently interferes with this regression of the male gland and may do so by delaying cell death.

Combinatorial activity of the *Wnt/int* genes with other factors could explain their seemingly contrasting functions in normal development and in mammary tumorigenesis. During development, the *int* genes are expressed in areas with either no or a low rate of proliferation and then probably act only as morphogens, inducing differentiation in embryonic tissue. Upon inadvertent activation by MMTV in mammary tissue, they may, in conjunction with other factors, act as strong mitogens. The combined activation of *Wnt*-1 and *int*-2 in some mammary tumors is therefore of particular interest (PETERS et al. 1986). The end result of the activity of many growth factors depends on the context in which these factors are present; transforming growth factor-β (TGF-β) plus FGF induces mesoderm in embryos; TGF-β plus TGF-α causes transformation of NRK cells (reviewed in SPORN and ROBERTS 1988). Of equal importance is the type of cell on which these factors act, some cells being stimulated and others inhibited by the very same peptide.

It is anticipated that new insight into the mechanism of action of the *Wnt/int* genes will be generated by inactivating the genes in the mouse germline and examining the phenotype of the resulting mutant mice. The possibility to manipulate embryonic stem cells in vitro to introduce mutations at will by homologous recombination has been explored for both *Wnt*-1 and *int*-2. The *Wnt*-1 gene has also been successfully inactivated in the germ line, with dramatic consequences on development (see Sect. 2.3), whereas *int*-2 mutations have been obtained at a relatively high frequency in ES cells but not yet in the germ line (MANSOUR et al. 1988; CAPECCHI 1989). Once obtained, such animals will be invaluable in examining the tasks of these intriguing genes in the developing animal, and, by extrapolation, in the process of mammary tumorigenesis.

Acknowledgements. I thank Drs. C. Dickson, G. Peters, J. Papkoff, R. Callahan, A. Brown, H. Varmus, and A. McMahon for preprints and permission to cite unpublished observations. I am especially indebted to Clive Dickson for providing the information on which Fig. 3 is based. Several colleagues provided useful comments on the manuscript.

References

Abraham JA, Whang JL, Tumolo A, Mergia A, Friedman J, Gospodarowicz D, Fiddes JC (1986) Human basic fibroblast growth factor: nucleotide sequence and genomic organization. EMBO J 5: 2523–2528

Acland P, Dixon M, Peters G, Dickson C (1990) Subcellular fate of the int-2 oncoprotein is determined by choice of initiation codon. Nature 343: 662–665

Adelaide J, Mattei M-G, Marics I, Raybaud F, Planche J, De Lapeyriere O, Birnbaum D (1988) Chromosomal localization of the hst oncogene and its co-amplification with the int.2 oncogene in a human melanoma. Oncogene 2: 413–416

Arheden K, Tommerup N, Mandahl N, Heim S, Winther S, Jensen OA, Prause JU, Mitelman F (1989) Amplification of the human putative oncogene INT1 in primary retinoblastoma tumors. Cytogenet Cell Genet 48: 174

Baker NE (1987) Molecular cloning of sequences from wingless, a segment polarity gene in Drosophila: the spatial distribution of a transcript in embryos. EMBO J 6: 1765–1773

Bouchard L, Lamarre L, Tremblay PJ, Jolicoeur P (1989) Stochastic appearance of mammary tumors in transgenic mice carrying the MMTV/c-neu oncogene. Cell 57: 931–936

Bradley RS, Brown AM (1990) The proto-oncogene int-1 encodes a secreted protein associated with the extracellular matrix. EMBO J 9: 1569–1575

Brookes S, Placzek M, Moore R, Dixon C, Peters G (1986) Insertion elements and transitions in cloned mouse mammary tumour virus DNA: further delineation of the poison sequences. Nucleic Acids Res 14: 8231–8245

Brookes S, Smith R, Casey G, Dickson C, Peters G, (1989a) Sequence organization of the human *int*-2 gene and its expression in tetratocarcinoma cells. Oncogene 4: 429–436

Brookes S, Smits R, Thurlow J, Dickson C, Peters G (1989b) The mouse homologue of hst/k-FGF: sequence, genome organisation and location relative to *int*-2. Nucleic Acids Res 17: 4037–4045

Brown AMC, Wildin RS, Prendergast TJ, Varmus HE (1986) A retrovirus vector expressing the putative mammary oncogene *int*-1 causes partial transformation of a mammary epithelial cell line. Cell 46: 1001–1009

Brown AMC, Papkoff J, Fung YKT, Shackleford GM, Varmus HE (1987) Identification of protein products encoded by the proto-oncogene *int*-1. Mol Cell Biol 7: 3971–3977

Buchberg AM, Brownell E, Nagata S, Jenkins NA, Copeland NG (1989) A comprehensive genetic map of murine chromosome 11 reveals extensive linkage conservation between mouse and human. Genetic 122: 153–161

Capecchi MR (1989) The new mouse genetics: altering the genome by gene by targeting. Trends Genet 5: 70–76

Chan AM-L, Hilkens J, Kroezen V, Scambler P, Williamson R, Cooper CS (1989) Molecular cloning and localization to chromosome 6 of the mouse irp gene. Somat Cell Mol Genet 15: 555–562

Davis CA, Joyner AL (1988) Expression patterns of the homeo box-containing genes En-1 and En-2 and the proto-oncogene *int*-1 diverge during mouse development. Genes Dev 2: 1736–1744

Delli-Bovi P, Curatola AM, Kern FG, Greco A, Ittmann M, Basilico C (1987) An oncogenes isolated by transfection of Kaposi's sarcoma DNA encodes a growth factor that is a member of the FGF family. Cell 50: 729–737

Delli-Bovi P, Curatola AM, Newman KM, Sato Y, Moscatelli D, Hewick RM, Rifkin DB, Basilico C (1988) Processing, secretion, and biological properties of a novel growth factor of the fibroblast growth factor family with oncogenic potential potential. Mol Cell Biol 8: 2933–2941

Dickson C, Peters G (1987) Potential oncogene product related to growth factors. Nature 326: 833

Dickson C, Smith R, Brookes S, Peters G (1984) Tumorigenesis by mouse mammary tumor virus: proviral activation of a cellular gene in the common integration region *int*-2. Cell 37: 529–536

Dickson C, Smith R, Brookes S, Peters G (1990) Proviral insertions within the *int*-2 gene can generate multiple anomalous transcripts but leave the protein-coding domain intact. J Virol 64: 784–793

DiNardo S, Sher E, Heemskerk-Jongens J, Kassis JA, O'Farrell PH (1988) Two- tiered regualtion of spatially patterned engrailed gene expression during Drosophila embryogenesis. Nature 332: 604–609

Dixon M, Deed R, Acland P, Moore R, Whyte A, Peters G, Dickson C (1990) Detection and characterization of the FGF-related oncoprotein int-2. Mol Cell Biol 9: 4896–4902

Fantl V, Brookes S, Smith R, Casey G, Barnes D, Johnstone G, Peters G, Dickson C (1989) Characterization of the proto-oncogene int-2 and its potential for the diagnosis of human breast cancers. Oncogene (in press)

Finch PW, Rubin JS, Miki T, Ron D, Aaroson SA (1989) Human KGF is FGF-related with properties of a paracrine effector of cell growth. Science 245: 752–755

Fung YKT, Shackleford GM, Brown AMC, Sanders GS, Varmus HE (1985) Nucleotide sequence and expression in vitro of cDNA derived from mRNA of int-1, a provirally activated mouse mammary oncogene. Mol Cell Biol 5: 3337–3344

Gallahan D, Callahan R (1987) Mammary tumorigenesis in feral mice- identification of a new int locus in mouse mammary tumor virus (Czech II)-induced mammary tumors. J Virol 61: 66–74

Gallahan D, Kozak C, Callahan R (1987) A new common integration region (int-3) for mouse mammary tumor virus on mouse chromosome 17. J Virol 61: 218–220

Garcia M, Wellinger R, Vessaz A, Diggelmann H (1986) A new site of integration for mouse mammary tumor virus proviral DNA common to BALB/cf(C3H) mammary and kidney adenocarcinomas. EMBO J 5: 127–134

Gospodarowicz D, Ferrera N, Schweigerer L, Neufeld G (1987) Structural characterization and biological functions of fibroblast growth factor. Endoc Rev 8: 95–114

Gray DA, Jackson DP, Percy DH, Morris VL (1986) Activation of int-1 and int-2 loci in GRF mammary tumors. Virology 154: 271–278

Guenzburg WH, Salmons B (1986) Mouse mammary tumor virus mediated transfer and expression of neomycin resistance,to infected cultured cells. Virology 155: 236–248

Hilgers J, Bentvelzen P (1979) Interaction between viral and genetic factors in murine mammary cancer. Adv Cancer Res 26: 143–195

Huebner K, Ferrari AC, Delli Bovi P, Croce CM, Basilico C (1988) The FGF-related oncogene, K-FGF, maps to human chromosome region 11q13, possibly near int-2. Oncogene Res 3: 263–270

Ingham PW (1988) The molecular genetics of embryonic pattern formation in Drosophila. Nature 335: 25–34

Ingham PW, Baker NE, Martinez-Arias A (1988) Regulation of segment polarity genes in the Drosophila blastoderm by fushi tarazu and even skipped. Nature 331: 73–75

Jakobovits A, Shackleford GM, Varmus HE, Martin GR (1986) Two proto-oncogenes implicated in mammary carcinogenesis, int-1, and int-2, are independently regulated during mouse development. Proc Natl Acad Sci USA 83: 7806–7810

Jaye M, Howk R, Burgess W, Ricca GA, Chiu I, Ravera MW, O'Brien SJ, Modi WS, Maciag T, Drohan WN (1986) Human endothelial cell growth factor: cloning, nucleotide sequence, and chromosome localization. Science 233: 541–545

Jones-Villeneuve EMV, Rudnicke MA, Harris JF, McBurney MW (1983) Retinoic acid-induced neural differentiation of embryonal carcinoma cells. Mol Cell Biol 3: 2271–2279

Kamb A, Weir M, Rudy B, Varmus H, Kenyon C (1989) Identification of genes from pattern formation, tyrosine kinase, and potassium channel families by DNA amplification. Proc Natl Acad Sci USA 86: 4372–4376

Kennedy N, Knedlitschek G, Groner B, Hynes NE, Herrlich P, Michalides R, Van Ooyen AJJ (1982) Long terminal repeats of endogenous mouse mammary tumor virus contain a long open reading frame which extends into adjacent sequences. Nature 295: 622–624

Kimelman D, Kirschner M (1987) Synergistic induction of mesoderm by FGF and TGF-beta and the identification of an mRNA coding for FGF in the early Xenopus embryo. Cell 51: 869–877

Kratochwil K (1985) The importance of epithelial stromal interaction in mammary gland development. In: Rich MA, Hager JC, Taylor-Papadimitriou J (eds) Breast cancer: origins, detection, and treatment. Nijhoff Boston pp 1–12

Leder A, Pattengale PK, Kuo A, Stewart T, Leder P (1986) Consequences of widespread deregulation of the c-myc gene in transgenic mice: multiple neoplasms and normal development. Cell 45: 485–495

Lee PL, Johnson DE, Cousens LS, Fried VA, Williams LT (1989) Purification and complementary DNA cloning of a receptor for basic fibroblast growth factor. Science 245: 57–60

Lidereau R, Callahan R, Dickson C, Peters G, Escot C, Ali IU (1988) Amplification of the int-2 gene in primary human breast tumors. Oncogenes Res 2: 285–291

Majors J, Varmus HE (1983) Nucleotide sequencing of an apparent proviral copy of env mRNA defines determinants of expression of the mouse mammary tumor virus env gene. J Virol 47: 495–504

Mansour SL, Martin GR (1988) Four classes of mRNA are expressed from the mouse int-2 gene, a member of the FGF family. EMBO J 7: 2035 2041

Mansour SL, Thomas KR, Capecchi MR (1988) Disruption of the proto-oncogene int-2 in mouse embryo-derived stem cells: a general strategy for targeting mutations to non-selectable genes. Nature 336: 348–352

Marics I, Adelaide J, Raybaud F, Mattei M-G, Coulier C, Planche J, de Lapeyriere O, Birnbaum D (1989) Characterization of the HST-related FGF.6 gene, a new member of the fibroblast growth factor gene family. Oncogene 4: 335–340

Martinez Arias A, Baker NE, Ingham PW (1988) Role of segment polarity genes in the definition and maintenance of cell states in the Drosophila embryo. Development 103: 157–170

McMahon AP, Moon RT (1989a) Ectopic expression of the proto-oncogene int-1 in Xenopus embryos leads to duplication of the embryonic axis. Cell 58: 1075–1084

McMahon AP, Moon RT (1989b) int-1—a proto-oncogene involved in cell signaling. Development 107: 161–167

McMahon JA, McMahon AP (1989) Nucleotide sequence, chromosomal localization and developmental expression of the mouse int-1 related gene. Development 107: 643–650

Mester J, Wagenaar E, Sluyser M, Nusse R (1987) Activation of the int-1 and int-2 mammary oncogenes in hormone-dependent and independent mammary tumors of GR mice. J Virol 61: 1073–1078

Muller WJ, Lee SF, Dickson C, Peters G, Pattengale P, Leder P (1990) The int-2 gene product acts as an epithelial growth factor in transgenic mice. EMBO J 9: 907–913

Moore R, Casey G, Brookes S, Dixon M, Peters G, Dickson C (1986) Sequence, topography and protein coding potential of mouse int-2: a putative oncogene actived by mouse mammary tumor virus. EMBO J 5: 919–924

Moore R, Dixon M, Smith R, Peters G, Dickson C (1987) Complete nucleotide sequence of a milk-transmitted mouse mammary tumor virus: two frameshift suppression events are required for translation of gag and pol. J Virol 61: 480–490

Morata G, Lawrence PA (1977) The development off wingless, a hometic mutation of Drosophila. Dev Biol 56: 227–240

Morris DW, Barry PA, Bradshaw HJ, Cardiff RD (1990) Insertion mutation of the int-1 and int-2 loci by mouse mammary tumor virus in premalignant and malignant neoplasms from the GR mouse strain. J Virol J 64: 1794–1802

Muller WJ, Sinn E, Pattengale PK, Wallace R, Leder P (1988) Single-step induction of mammary adenocarcinoma in transgenic mice bearing the activated c-neu oncogene. Cell 54: 105–115

Morris DW, Bradshaw HD, Billy HT, Munn RJ, Cradiff RD (1989) Isolation of a pathogenic clone of mouse mammary tumor virus. J Virol 63: 148–158

Nakano Y, Guerrero I, Hildalgo A, Taylor A, Whittle JRS, Ingham PW (1989) The Drosophila segment polarity gene patched encodes a protein with multiple potential membrane spanning regions. Nature 341: 508–513

Noordermeer J, Meijlink F, Verrijzer P, Rijsewijk F, Destree O (1989) Isolation of the Xenopus homolog of int-1/wingless and expression during neurula stages of early development. Nucleic Acids Res 17: 11–18

Nusse R (1986) The activation of cellular oncogenes by retroviral insertion. Trends Genet 2: 244–247

Nusse R (1988a) The activation of cellular oncogenes by proviral insertion in murine mammary tumors. In: Lippman ME, Dickson R (eds) Breast Cancer: cellular and molecular biology. Kluwer, Boston; pp 283–306

Nusse R (1988b) The int genes in mammary tumorigenesis and in normal development. Trends Genet 4: 291–295

Nusse R, Varmus HE (1982) Many tumors induced by the mouse mammary tumor virus contain a provirus integrated in the same region of the host genome. Cell 31: 99–109

Nusse R, Van Ooyen A, Cox D, Fung YKT, Varmus HE (1984) Mode of proviral activation of a putative mammary oncogene (int-1) on mouse chromosome 15. Nature 307: 131–136

Nusse R, Theunissen J, Wagenaar E, Rajsewijk F, Gennissen A, Otte A, Schuuring E, Van Ooyen A (1990) The Wnt-1 (int-1) oncogene promoter and its mechanism of activation by insertion of proviral DNA of the mouse mammary tumor virus. Mol Cell Biol 10: 4170–4179

Papkoff J (1989) Inducible overexpression and secretion of int-1 protein. Mol Cell Biol 8: 3377–3384

Papkoff J, Schryver B (1990) Secreted *int*-1 protein is associated with the cell surface. Mol Cell Biol 10: 2723–2730

Papkoff J, Brown AMC, Varmus HE (1987) The *int*-1proto-oncogene products are glycoproteins that appear to enter the secretory pathway. Mol Cell Biol 7: 3978–3984

Paterno GD, Gillespie LL, Dixon MS, Slack JMW, Heath JK (1989) Mesoderm-inducing properties in INT-2 and kFGF: two oncogene-encoded growth factors related to FGF. Development 106: 79–83

Perrimon N, Mahowald AP (1987) Multiple functions of segment polarity genes in Drosophila. Dev Biol 119: 587–600

Perrimon N, Smouse D (1989) Multiple functions of a Drosophila homeotic gene, zeste-white 3, during segmentation and neurogenesis. Dev Biol 135: 287–305

Peters G, Brookes S, Smith R, Dickson C (1983) Tumorigenesis by mouse mammary tumor virus:Evidence for a common region for provirus integration in mammary turmors. Cell 33: 369–377

Peters G, Lee A, Dickson C (1984) Activation of cellular gene by mouse mammary tumor virus may occur early in tumor development. Nature 309: 273–275

Peters G, Lee AE, Dickson C (1986) Concerted activation of two potential proto-oncogenes in carcinomas induced by mouse mammary tumor virus. Nature 320: 628–631

Peters G, Brookes S, Placzek M, Schuermann M, Michalides R, Dickson C (1989a) A putative int domain for mouse mammary tumor virus on mouse chromosome 7 is a 5′ extension of *int*-2. Virol J 63: 1448–1450

Peters G, Brookes S, Smith R, Placzek M, Dickson C (1989b) The mouse homolog of the hst/k-FGF gene is adjacent to *int*-2 and is activated by proviral insertion in some virally induced mammary tumors. Proc Natl Acad Sci USA 86: 5678–5682

Rijsewijk FA, Van Lohuizen M, Van Ooyen A & Nusse R (1986). Construction of a retroviral cDNA version of the *int*-1 mammary oncogene and its expression in vitro. Nucleic Acids Res 14: 693–702

Riggleman B, Wieschaus E, Schedl P (1989) Molecular analysis of the armadillo locus: uniformly distributed transcripts and a protein with novel internal repeats are associated with a Drosophila segment polarity gene. Genes Dev 3: 96–113

Rijsewijk F, Schuermann M, Wagenaar E, Parren P, Weigel D, Nusse R (1987a) The Drosophila homologue of the mammary oncogene *int*-1 is identical to the segment polarity gene wingless. Cell 50: 649–657

Rijsewijk F, Van Deemter L, Wagenaar E, Sonnenberg A, Nusse R (1987b) Transfection of the *int*-1 mammary oncogene in cuboidal RAC mammary cell line results in morphological transformation and tumorigenicity. EMBO J 6: 127–131

Roelink H, Wagenaar E, Lopes da Silva S, Nusse R (1990) *Wnt*-3, a gene activated by proviral insertion in mouse mammary tumors, is homologous to *int*/Wnt-1 and normally expressed in mouse embryos and adult brain. Proc Natl Acad Sci USA 87: 4519–4523

Rogelj S, Weinberg RA, Fanning P, Klagsbrun M (1988) Basic fibroblast growth factor fused to a signal peptide transforms cells. Nature 331: 173–175

Sakamoto H, Yoshida T, Nakakuki M, Odagiri H, Miyagawa K, Sugimura T, Terada M (1988) Cloned hst gene from normal human leukocyte DNA transformations NIH3T3 cells. Biochem Biophys Res Commun 151: 965–972

Schuermann M, Michalides R (1987) A rare common integration site of proviruses of the mouse mammary tumor virus in P-type mammary tumors of mouse strain GR. Virology 156: 229–237

Schuuring E, van Deemter E, Roelink H, Nusse R (1989) Expression of the *int* 1 proto oncogene during differentiation of P19 embryonal carcinoma cells. Mol Cell Biol 9: 1357–1361

Schuuring E, Van der Leede BJ, Willems R, Daams JH, Van der Valk M, Van der Vijver MG, Van Leeuwen F, Sonnenberg A, Nusse R (1990) Differentiation dependent expression of provirus activated *int*-1 oncogene in clonal cell lines derived from a mouse mammary tumor. Oncogene 5: 459–465

Shackleford GM, Varmus HE (1987) Expression of the proto-oncogene *int*-1 is restricted to postmeiotic male germ cells and the neural tube of mid-gestational embryos. Cell 50: 89–95

Shackleford GM, Varmus HE (1988) Construction of a clonable, infectious, and tumorigenic mouse mammary tumor virus provirus and a derivative genetic vector. Proc Natl Acad Sci USA 85: 9655–9659

Slack JMW, Darlington BG, Heath JK, Godsave SF (1987) Mesoderm induction in early xenopus embryos by heparin-binding growth factor. Nature 326: 197–200

Smith R, Peters G, Dickson C (1988) Multiple RNAs expressed from the *int*-2 gene in mouse embryonal carcinoma cell lines encode a protein with homology to fibroblast growth factors. EMBO J 7: 1013–1022

Sonnenberg A, Van Balen P, Hilgers J, Schuuring E, Nusse R (1987) Oncogene expression during progression of mouse mammary tumor cells; activity of a proviral enhancer and the resulting expression of int-2 is influenced by the state of differentiation. EMBO J 6: 121–125

Sporn MB, Roberts AB (1988) Peptide growth factors are multifunctional. Nature 332: 217–219

St-Arnaud R, Craig J, McBurney MW, Papkoff J (1989) The int-1 proto-oncogene is transcriptionally activated during neuroectodermal differentiation of P19 mouse embryonal carcinoma cells. Oncogene 4: 1077–1080

Taira M, Yoshida T, Miyagawa K, Sakamoto H, Terada M, Sugimura T (1987) cDNA sequence of human transforming gene hst and identification of the coding sequence required for transforming activity. Proc Natl Acad Sci USA 84: 2980–2984

Tsuda H, Hirohashi S, Shimosato Y, Hirota T, Tsugane S, Yamamoto H, Miyajima N, Toyoshima K, Yamamoto T, Yokota J, Yoshida T, Sakamoto H, Terada M, Sugimura T (1989) Correlation between long-term survival in breast cancer patients and amplification of two putative oncogene-coamplification units: hst-1/int-2 and c-erbB-2/ear-1 Cancer Res 49: 3104–3108

Tsukamoto AS, Grosschedl R, Guzman RC, Parslow T, Varmus HE (1988) Expression of the int-1 gene in transgenic mice is associated with mammary gland hyperplasia and adenocarcinomas in male and female mice. Cell 55: 619–625

Tsutsumi M, Sakamoto H, Yoshida T, Kakizoe T, Koiso K, Sugimura I, Terada M (1988) Coamplification of the hst-1 and int-2 genes in human cancers. Jpn J Cancer Res 79: 428–432

Uzvolgyi E, Kiss I, Pitt A, Arsenian S, Ingvarsson S, Udvardy A, Hamada M, Klein G, Sumegi J (1988) Drosophila homolog of the murine int-1 protooncogene. Proc Natl Acad Sci USA 85: 3034–3038

Van den Heuvel M, Nusse R, Johnston P, & Lawrence PA (1989). Distribution of the wingless gene product in Drosophila embryos:a protein involved in cell–cell communication. Cell 59: 739–49

Van Klaveren P, Bentvelzen P (1988) Transactivating potential of the 3′ open reading frame of murine mammary tumor virus. J Virol 62: 4410–4413

Van Ooyen A, Nusse R (1984) Structure and nucleotide sequence of the putative mammary oncogene int-1 proviral insertions leave the protein-encoding domain intact. Cell 39: 233–240

Van Ooyen A, Kwee V, Nusse R (1985) The nucleotide sequence of the human int-1 mammary oncogene; evolutionary conservation of coding and noncoding sequences. EMBO J 4: 2905–2909

Van't Veer LJ, Geurts van Kessel A, Van Heerikhuizen H, Van Ooyen A, Nusse R (1984) Molecular cloning and chromosomal assignment of the human homolog of int-1, a mouse gene implicated in mammary tumorigenesis. Mol Cell Biol 4: 2532–2534

Varley JM, Walker RA, Casey G, Brammar WJ (1988) A common alteration to the int-2 proto-oncogene in DNA from primary breast carcinomas. Oncogene 3: 87–91

Varmus HE (1984) The molecular genetics of cellular oncogenes. Annu Rev Genet 18: 553–612

Wainwright BJ, Scambler PJ, Stainer P, Watson EK, Bell G, Wicking C, Estivill X, Courtney M, Boue A, Pedersen PS, Williamson R, Farrall M(1988) Isolation of a human gene with protein sequence similarity to human and murine int-1 and the Drosophila segment polarity mutant wingless. EMBO J 7: 1743–1748

Wasylyk B, Wasylyk C, Augereau P, Chambon P (1983) The SV40 72 bp repeat preferentially potentiates transcription starting from proximal natural or substitute promoter elements. Cell 32: 503–514

Wilkinson DG, Bailes JA, McMahon AP (1987) Expression of the proto-oncogene int-1 is restricted to specific neural cells in the developing mouse embryo. Cell 50: 79–88

Wilkinson DG, Peters G, Dickson C, McMahon AP (1988) Expression of the FGF-related proto-oncogene int-2 during gastrulation and neurulation in the mouse. EMBO J 7: 691–695

Wilkinson DG, Bhatt S, McMahon AP (1989) Expression Pattern of the FGF-related proto-oncogene int-2 suggests multiple roles in fetal development. Development 105: 131–136

Yoshida MC, Wada M, Satoh H, Yoshida T, Sakamoto H, Miyagawa K, Yokota J, Koda T, Kakinuma M, Sagimura T, Terada M (1988) Human HST1 (HSTF1) gene maps to chromosome band 11q13 and coamplifies with the INT2 gene in human cancer. Proc Natl Acad Sci USA 85: 4861–4864

Yoshida T, Miyagawa K, Odagiri H, Sakamoto H, Little PFR, Terada M, Sugimura T (1987) Genomic sequence of hst, a transforming gene encoding a protein homologous to fibroblast growth factors and the int-2-encoded protein. Proc Natl Acad Sci USA 84: 7305–7309

Zhan X, Bates B, HU X, Goldfarb M (1988) The human FGF-5 oncogene encodes a novel protein related to fibroblast growth factors. Mol Cell Biol 8: 3487–3495

Zhou DJ, Casey G, Cline MJ (1988) Amplification of human int-2 in breast cancers and squamous carcinomas. Oncogene 2: 279–282

Feline Leukaemia Virus: Generation of Pathogenic and Oncogenic Variants

J. C. Neil, R. Fulton, M. Rigby, and M. Stewart

1	Introduction	67
1.1	Origins of FeLV	68
1.2	Endogenous FeLV-Related Sequences	68
2	Evolution of Pathogenic and Oncogenic Variants of FeLV	69
2.1	Envelope Gene Variants	69
2.1.1	FeLV Subgroups	69
2.1.2	FeLV-A: The Prevalent Form of FeLV	70
2.1.3	FeLV-B: Feline Homologue of Murine Mink Cell Focus-Forming Viruses	70
2.1.4	FeLV-C and Erythroid Hypoplasia	73
2.1.5	FeLV-Feline Acquired Immunodeficiency Syndrome	74
2.1.6	Envelope Sequence of enFeLV: Clues to the Targets of the Feline Immune Response	75
2.2	FeLV LTR Variation	76
2.2.1	Enhancer Duplication	76
2.2.2	Enhancer Duplications and Insertional Mutagenesis	78
2.2.3	Other LTR Mutations	78
2.3	FeLV Recombinants Carrying Host Cell Genes	79
2.3.1	FeLV Transduction of Cellular Genes: A Feature of Naturally Occurring Tumours	79
2.3.2	Mechanism of Transduction: Remaining Problems	80
2.3.3	Recombinational, Junctions in FeLV	81
2.3.4	Recombination Sites in c-myc	83
3	Conclusions	85
3.1	Lessons from a Natural Leukaemogenesis Model: Comparison of the Feline and Murine C-Type Retroviruses	85
3.2	Prospects	88
	References	88

1 Introduction

Although feline leukaemia virus (FeLV) is a common and potent pathogen of the domestic cat, neoplastic disease is a relatively rare outcome of the virus-host interaction and generally occurs after a long latent period (HARDY et al. 1976; JARRETT 1984). For these reasons, it has been customary to classify FeLV as a chronic leukaemogenic retrovirus and to place the acute transforming C-type retroviruses in a separate subclass. In our view, this classification has been

Beatson Institute for Cancer Research, Bearsden, Glasgow G61 1BD, UK unhelpful since it obscures the fact that acute transforming retroviruses invariably arise from the chronic leukaemogenic variety. Moreover, there is mounting evidence that terminal diseases which develop in FeLV-infected cats are induced by variant genomes that arise de novo. In this review, we discuss the origins of FeLV variants and briefly consider their roles in disease.

The spectrum of neoplastic diseases observed in FeLV-infected cats has been reviewed elsewhere (HARDY et al. 1980; JARRETT 1984; TEICH et al. 1982). The most common are lymphoid tumours of the T-cell series, which arise as primary thymic tumours or as multicentric tumours affecting multiple lymph nodes. Myeloid and erythroid leukaemias also occur, though with lower frequency. Alimentary lymphoid tumours of presumptive B-cell origin are also commonly observed tumours of cats, but these are generally devoid of infectious FeLV and occur predominantly in older animals. Multicentric fibrosarcomas are relatively rare, but these have been of considerable interest because they frequently yield recombinant feline sarcoma viruses.

In contrast to neoplastic disease, immunosuppression by FeLV is a common clinical problem and probably reflects an intrinsic property of FeLV which is essential for its survival as a horizontally transmitted agent. Immune suppression may also be an important early step in the neoplastic process. However, highly immunosuppressive variants are probably self-limiting outside of the laboratory, as appears to be the case for the highly oncogenic variants (ONIONS et al. 1987). Whilst a knowledge of virus variants is essential to understand the pathogenesis of FeLV infection, such variants have thus far proved incompetent for horizontal transmission and therefore appear to be of little epidemiological significance.

1.1 Origins of FeLV

FeLV is a C-type retrovirus with the typical genetic structure: 5'-LTR-*gag-pol-env*-LTR-3'. FeLV appears to have evolved from an ancestral rodent virus (BENVENISTE et al. 1975) and is remarkably closely related in genetic structure and primary sequence to the leukaemogenic C-type retroviruses of the laboratory mouse (MuLV). Although there are many similarities between FeLV and the more widely studied MuLVs, the feline virus offers the unique perspective of a successful pathogen which thrives in a natural, outbred population, and comparisons between the two viral systems are frequently instructive.

1.2 Endogenous FeLV-Related Sequences

The domestic cat and a group of small related Felidae harbour endogenous sequences related to FeLV (enFeLV) (BENVENISTE et al. 1975; BENVENISTE and TODARO 1975; OKABE et al. 1976). The divergence between exogenous and endogenous FeLV was first shown by subtractive cDNA probes which indicated that a portion of the exogenous FeLV genome was missing from the enFeLV sequence (OKABE et al. 1978). Subsequently, it was found that the majority of the enFeLV sequences represent essentially full-length copies although a significant minority is truncated (SOE et al. 1985). Also, at least two domains of the enFeLV genome diverge significantly from exogenous FeLV: the U3 region of the long terminal repeat (LTR) (CASEY et al. 1981); and a portion of the *env* gene (STEWART et al. 1986b). A number of the enFeLV sequences have been molecularly cloned (SOE et al. 1983, 1985), and partial sequencing has confirmed the conclusions drawn from hybridization analyses (BERRY et al. 1988; KUMAR et al. 1989). Even the apparently full-length enFeLV proviruses do not appear to encode infectious viruses (J. MULLINS, personal communication) and expression of enFeLV appears to be limited to subgenomic transcripts (BUSCH et al. 1983; NIMAN et al. 1977a, b, 1980). Despite their apparent innocuity, enFeLV can contribute to FeLV pathogenesis, as discussed in Sect. 2.

2 Evolution of Pathogenic and Oncogenic Variants of FeLV

2.1 Envelope Gene Variants

2.1.1 FeLV Subgroups

Replication-competent FeLV isolates can be classified into three subgroups, A, B and C (JARETT et al. 1973; SARMA and LOG 1973). The subgroup phenotype is determined by the viral envelope and is thought to reflect the recognition of three distinct host cell receptors by FeLV. Interference to superinfection by a virus of the same subgroup would then arise from the blockade of receptors by the envelope glycoproteins expressed from a resident provirus. Table 1 summarizes the biological properties of FeLV-A, B and C, which have been considered in detail in other reviews (JARRETT 1984).

Table 1. FeLV subgroups: occurrence in natural isolates and biological properties

Subgroup	Occurrence in field isolates (%)	Origin	Biological properties
A	100	Infectious spread	Often low pathogenicity Efficiently transmitted Generally ecotropic
B	33–67[a]	Recombinant between FeLV-A and enFeLV	Heterogeneous Some replication-defective Some accelerate leukaemia
		Spread along with FeLV-A	Expanded host range
C	<1	Mutation from FeLV-A?	Induce erythroid hypoplasia Expanded host range

[a] Higher figure recorded in leukaemic cats, lower figure in FeLV infected healthy cats

2.1.2 FeLV-A: The Prevalent Form of FeLV

The distribution of FeLV subgroups in the cat population is profoundly biased in that every isolate contains a subgroup A virus (JARRETT et al. 1978a). The commonly isolated FeLV-A viruses (e.g., A/Glasgow-1, 1161E) are generally of low to moderate pathogenicity, but efficiently horizontally transmitted.

Figure 1 shows an alignment of the gp 70^{env} protein sequences predicted from DNA sequence analysis of various FeLV *env* genes. Four complete FeLV-A *env* sequences have been determined from three independent isolates, and these show very high homology despite their geographically separate isolation (DONAHUE et al. 1988; STEWART et al. 1986b). Furthermore, serological analyses indicate that FeLV-A isolates are antigenically strongly conserved and behave in an essentially monotypic fashion in virus neutralization assays (RUSSELL and JARRETT 1978a). FeLV-A *env* conservation presumably reflects strong constraints on the sequence for efficient horizontal transmission (JARRETT and RUSSELL 1978) and bears out the suggestion that protective vaccines against FeLV infection should be based on FeLV-A, since they are the primary infectious viruses and hence the ultimate source of all the pathogenic variants (JARRETT 1984).

2.1.3 FeLV-B: Feline Homologue
of Murine Mink Cell Focus-Forming Viruses

FeLV-B viruses are relatively common (present in 40%–50% of all isolates), but are overrepresented in leukaemic cats (> 60%). For FeLV-B, as for FeLV-C, apparent in vivo dependence on FeLV-A is not generally due to simple replication defectiveness, since the subgroup variants can often be separated and grown to similar titres in cultured feline fibroblast cells.

A series of interrelated observations led to the conclusion that FeLV-B viruses arise by recombination between FeLV-A and enFeLV *env* sequences. First, sequence analysis of an FeLV-B *env* gene showed relatedness to murine mink cell focus-forming (MCF) viruses (ELDER and MULLINS 1983), which are themselves formed by recombination of exogenous and endogenous proviral sequences (KHAN 1984). Second, sequence analysis of FeLV A *env* genes (LUCIW et al. 1985; STEWART et al. 1986b) revealed that the FeLV-B 5' *env* domain with closest homology to MCF was not well matched with FeLV-A viruses. Furthermore, DNA hybridization probes, constructed from the most divergent region of the *env* genes of FeLV-A and FeLV-B, provided evidence that FeLV-B viruses consistently retain a 5' portion related to enFeLV (STEWART et al. 1986b). Finally, direct sequence comparison of FeLV-A, FeLV-B, and enFeLV gave detailed confirmation of the recombination process (J. MULLINS, personal communication; KUMAR et al. 1989; NEIL et al. 1987). It can be seen clearly from Fig. 1 that the FeLV-B viruses have all acquired variable regions (vr) 1–4 from enFeLV. The minimal enFeLV sequence content which defines a viable subgroup B virus has yet to be defined.

```
          <Leader                                   <Start
A-Glasgow MESPTHPKPSKDKTLSWNLAFLVGILFTIDIGMANPSPHQIYNVTWVITNVQTNTQANATSMLGTL
A-G1(L)   .............................................................
A-F3      ...................................V.........................
A-1161E   ...................................V.........................
?-61C     ...................................V.........P.M..............
C-Sarma   .............FP......V.......Q..M.........V...............SR..
B-GA      ....................V.....................V......T...LV.G.K...
B-ST      ..G.........F..D.MI...V.LRL.V.............T...LV.G.K...
B-Rickard .............................................T...LV.G.K...
en-CFE-6  ..G.........F..D.II...V.LRL.A.............V......T...LV.GIK...
en-CFE-16 ..G.........F..D.II...V.LRL.V.............V......T...LV.G.K...


A-Glasgow TDAYPTLHVDLCDLVGDTWEPIVLNPTNVKHGARYSSSKYGCKTTDRKKQQQTYPFYVCPGHA
A-G1(L)   .............................................................
A-F3      .....................N.........D.............................
A-1161E   ..V......................S..............P....................
?-1161C   ..V....................M..S..........G.PP....................
C-Sarma   ........Y.............APD.    RSW......TH....................
B-GA      ...F..MYF....II.N..N.         SDQEPFPG...DQPM.RW..RNT........
B-ST      ...F..MYF....II.N..N.         SDQEPFPG...DHPM.RW..RNT........
B-Rickard ...F..MYF....II.N..N.         SDQEPFPG...DQPM.RW..RNT........
en-CFE-6  ..TF..IYF....II.N..N.         SDQEPFPG...DQPM.RW..RNTA.......
en-CFE-16 ...F...YF....II.N..N.         SGQEPFPG...DQPM.RW..RNTA.......
                                      ─────────────────────────────
                                                   vr1

A-Glasgow PSLGPKGTHCGGAQDGFCAAWGCETTGETWWKPTSSWDYITVKRG
A-G1(L)   ...................................A.........
A-F3      ...................................A....S.....
A-1161E   ...................................A....S.....
?-1161C   ...................................A....S.....
C-Sarma   ..M.....Y..........................A.........
B-GA      NRKQ      ...P.....V.........Y.R...........K.VTQGIYQCSGGGWCGPCYDKAVH
B-ST      NRKQ      ...P.....V.........Y.R...........K.VTQGIYQCSGGGWCGPCYDKAVH
B-Rickard NRKQ      ...P.....V.........Y.R...........K.VTQGIYQCSGGGWCGPCYDKAVH
en-CFE-6  NRKQ      ...P.....V.........Y.............K.VTQGIYQCNGGGWCGPCYDKAVH
en-CFE-16 NRKQ      ...P.....V.........Y.............K.VTQGIYQCSGGGWCGPCYDKAVH

          ─────             ─────                   ──────────────────
           vr2               vr3

A-Glasgow SSQDNSCE GKCNPLVLQFTQKGRQASWDGPKMWGLRLYRTGYDPIALFTVSRQVSTITPPQAMGPNL
A-G1(L)   ........ ...........................................................
A-F3      ........ ........I..................................................
A-1161E   .....N.. ........I.....K............................................
?-1161C   .....N.. ........I.....K............................................
C-Sarma   .N.....K ...............R.........S.........S.....M.................
B-GA      ..TTGAS.G.R....I........T.....S.........S.....S.....M.................
B-ST      ..ITGAS.G.R....I........T.....S.........S.....S.....M.................
B-Rickard ..TTGAG.G.R....I........T.....S.........S.....S.....M.................
en-CFE-6  ..TTGAS.G.R....I........T.....S.........S.....S.....M...............P
en-CFE-16 ..TTGAS.G.R....I........T.....S.........S.....S.....M..............DP

          ────────

A-Glasgow VLPDQKPPSRQSQTGSKVATQRPQTNES          APRSVAPTTMGPKRIGTGDRLINLVQGTY
A-G1(L)   ...........................          ........................R.......
A-F3      ....................L.....          .S.......VV..................
A-1161E   ...........................          .........V...................
?-1161C   ...........................          .........V...................
C-Sarma   ...........K...T.....ITS.........T....SA.....
B-GA      ...........IE.R.TPHHS.G.GGTPGITLVNASI..L.TPV.PAS.............
B-ST      ...........IE.R.TPHHS.G.GGTPGITLVNASI..L.TPV.PAS.......N.....
B-Rickard ...........IE.R.TPHHS.G.GGTPGITLVNASI..L.TPV.PAS.......N.....
en-CFE-6  ...........IE.R.IPHHP.G.GGTPGITLVNASI..L.TPV.PAS.......N.....
en-CFE-16 ...........IE.R.IPHPS#(term)
                              ─────────────────────────────────
                                           vr4
```

Fig. 1 (Continued)

```
A-Glasgow  LALNATDPNKTKDCWLCLVSRPPYYEGIAILGNYSNOTNPPPSCLSIPQHKLTISEVSGQGLCIGTVP
A-G1(L)    ..................................................................
A-F3       .........................................................T........
A-1161E    ..................................................................
?-1161C    ...........................................................I..P....
C-Sarma    ...............................V.........................T.........
B-GA       ..........R.......................................................
B-ST       ...V.N...........................V..................D..............
B-Rickard  ...V.N...........................V..................D..............
en-CFE-6   .T..V.N..........................V..................V.............A...


A-Glasgow  KTHQALCNKTQQGHTGAH         YLAAPNGTYWACNTGLTPCISMAVLNWTSDFCVLIELWPRVTY
A-G1(L)    ........E.........         ..........................................
A-F3       ........E.........         ..........................................
A-1161E    ..................         ..........................................
?-1161C    ..........H......DYLTAPR    ..........................................
C-Sarma    ......K...K..K.T..         ..........................................
B-GA       ........E.........         ..........................................
B-ST       ......K...K..K.T..         .....S.........I..........................
B-Rickard  ......K...K..K.T..         ...V.S....................................
en.CFE.6   ......K...R.T.....         ..V.....................................T.....I..


                    vr5

A-Glasgow  HQPEYVYTHFAKAVRFRR
A-G1(L)    ..................
A-F3       ..................
A-1161E    ..................
?-1161C    ..........G.......
C-Sarma    .....I....D.......
B-GA       .............A....
B-ST       ..........D.TV.L..
B-Rickard  ..................
en-CFE-6   .E...I.S..ENKP..K.
```

Fig. 1. Alignment of the SU (gp70^{env}) coding sequences of exogenous and endogenous FeLV proviruses. The FeLV-A/Glasgow-1 sequence is shown in full (STEWART et al. 1986b). For the other sequences, amino acid identities with A/Glasgow-1 are indicated by *dots*. The sequences shown are: an independent clone of FeLV-A/Glasgow-1 or G1(L) (LUCIW et al. 1985), FeLV-A/1161E and FeLV-A/F3A (DONAHUE et al. 1988), FeLV-C/Sarma (RIEDEL et al. 1986), FeLV-B/Gardner-Arnstein (ELDER and MULLINS 1983), FeLV-B/Snyder-Theilen (NUNBERG et al. 1984b), FeLV-B/Rickard (ELDER and MULLINS 1985), and enFeLV clones CFE-6 and CFE-16 (KUMAR et al. 1989). Variable regions as discussed in the text are delineated underneath the sequence as *vr* 1–5

Recombination resulting in the generation of FeLV-B can be reproduced readily in vitro after DNA transfection of feline fibroblast cells with FeLV-A proviral DNA (OVERBAUGH et al. 1988b). The mechanism by which this occurs has not yet been analyzed in detail. It seems most likely that recombination follows copackaging of enFeLV transcripts in FeLV-A by virtue of conservation of the presumptive packaging signal (BERRY et al. 1988). However, passage of FeLV-A in feline fibroblasts does not lead to recombination at the frequency reported after DNA transfection, and it should be considered that the transfection procedure may favour other non-retroviral recombination mechanisms (JASIN and BERG 1988).

The prevalence of FeLV-B viruses in natural isolates is not unexpected from the relative ease with which they are generated by recombination. However, FeLV-B viruses are more common in leukaemic cats than in FeLV-A infected healthy cats. Two possible interpretations have been advanced to explain these observations: either FeLV-B contributes to the leukaemogenic process or, alternatively, leukaemic cats have longer standing infections, allowing more opportunity for FeLV-B recombinants to arise (JARRETT et al. 1978a).

Few studies have addressed the pathogenesis of subgroup B viruses in the absence of FeLV-A. Direct inoculation of the FeLV-B/Sarma isolate into newborn cats overcame the normal requirement for FeLV-A, although persistent infection was established in only 15%–20%. However, a number of these cats developed thymic lymphosarcoma with no evidence of involvement of FeLV-A. Attempts to transmit virus from the FeLV-B-infected cats were unsuccessful (JARRETT et al. 1978b; JARRETT and RUSSELL 1978). An example of a FeLV-B recombinant associated with nonthymic leukaemogenesis is the FeLV-GM1 isolate, which induces predominantly myeloid leukaemias. The replication-defective FeLV-B/GM1 component potentiates establishment of viraemia and disease development by its weakly pathogenic FeLV-A helper virus (TZAVARAS et al. 1990). In conclusion, FeLV-B viruses can contribute to FeLV pathogenesis but in a complex and isolate-specific manner. In this and other respects, subgroup B viruses are analogous to murine MCF viruses, which also form a distinct interference subgroup (REIN 1982) and are generated by recombination with endogenous proviruses from which they acquire at least the 5′ portion of *env* (KHAN 1984).

2.1.4 FeLV-C and Erythroid Hypoplasia

FeLV-C viruses are rare and isolation is invariably from anaemic cats. Only one complete FeLV-C sequence is known at present (LUCIW et al. 1985; RIEDEL et al. 1986), but it is evident that FeLV-C/Sarma is much more closely related to FeLV-A than to FeLV-B. The sequences are colinear except in vr 1, where a three-codon deletion is required to align FeLV-C with A. Restriction fragment exchange experiments have shown that an 886-base-pair fragment spanning the *pol–env* junction confers host range and pathogenic properties of FeLV-C/Sarma on FeLV-A/1161E (RIEDEL et al. 1988). More recently, the crucial determinants of FeLV-C subgroup specificity have been narrowed down to the vr 1 domain itself (see Fig. 1; M. RIGBY, J. ROJKO, and J. NEIL, unpublished results). Sequence analysis of further FeLV-C isolates by polymerase chain reaction (PCR) amplification shows that each is altered at vr 1 relative to FeLV-A, and, while each precise structure is unique (M. RIGBY and J. NEIL, unpublished work; J. MULLINS, personal communication), they share the property of recognizing the subgroup C host cell receptor. An FeLV-A *env* probe derived from a wider domain encompassing vr 1 hybridizes to four independent FeLV-C isolates but is not closely related to any endogenous feline sequence (STEWART et al. 1986b). It appears, therefore, that the crucial vr 1 changes yielding subgroup C viruses are most likely mutational rather than recombinational.

An interesting feature of the FeLV-C/Sarma gp 70 sequence is that it resembles enFeLV rather than FeLV-A in vr 5 and at several other amino acid positions in the C-terminal half of gp 70. This provides a possible rationale for an earlier observation that the pattern of neutralization of FeLV-C/Sarma by cat sera correlated with that of several FeLV-B isolates rather than the essentially monotypic FeLV-A. Another FeLV-C isolate (FS246) behaved almost identically to FeLV-A in this respect (RUSSETT and JARRETT 1978a), showing that this property was distinguishable from virus subgroup or receptor specificity. FeLV-C/Sarma was isolated from a complex mixture of subgroups (FeLV-ABC/KT) produced by the FL74 tumour cell line (KAWAKAMI et al. 1967; SARMA et al. 1975) and may have acquired subgroup B/enFeLV-related epitopes as a result.

A particularly strong correlation exists between the presence of subgroup C viruses and acute hypoplasia of the erythroid series. Although rare, these variants have been isolated only from anaemic cats. Inoculation of several independent FeLV-C isolates into newborn cats reproduced the same characteristic disease (MACKEY et al. 1975; ONIONS et al. 1982) as did mixed virus stocks containing FeLV-C (HOOVER et al. 1974). The disease is characterized by the failure of erythroid precursor development past the BFU-E (ONIONS et al. 1982; TESTA et al. 1983) or CFU-E stage (ABKOWITZ et al. 1987a). The disease mechanism remains obscure, although it has been reported that erythroid cell progenitors are sensitized to complement-mediated lysis (ABKOWITZ et al. 1987b).

2.1.5 FeLV-Feline Acquired Immunodeficiency Syndrome

A further example in which a minimal change in the *env* sequence appears to have a profound effect on FeLV pathogenicity is in the feline acquired immunodeficiency syndrome (FAIDS) variant, FeLV-1161C. The subgroup phenotype of the FAIDS variant has not been reported, although the *env* gene of its associated subgroup A helper virus (1161E) blocks its entry, suggesting that it can recognize the A receptor (J. MULLINS, personal communication). Mapping of the determinants of the rapid FAIDS disease by construction of virus chimaeras between 1161C and the minimally pathogenic FeLV-1161E has indicated that the *env* gene carries a primary disease determinant, although mutational changes in the LTR may also enhance disease expression. The *env* gene of 1161C differs from that of 1161E by a six-codon deletion at vr 1 and a similar sized insertion at vr 5 as well as a number of points mutations (OVERBAUGH et al. 1988a).

The 1161C variant was isolated as a replication-defective virus, but competent forms have been generated by replacing the *gag–pol* region with that of the associated helper virus 1161E. Like the original virus complex, the reconstructed virus induced an acute immunosuppressive disease, with death occurring rapidly due to inflammatory changes in intestinal tissues. At the gross level, it appears that 1161E and 1161C infect the same range of cells and that only the outcome of infection differs. The 1161C variant *env* gene product is aberrantly processed (POSS et al. 1989), and cells expressing this gene product remain susceptible to superinfection with 1161E (J. MULLINS personal communication). These observations provide a possible basis for the cytopathic properties of the FAIDS isolate and the accumulation of unintegrated viral DNA in affected tissues (MULLINS et al. 1986). The prevalence of acute immunosuppressive variants such as 1161C is unknown at present.

2.1.6 Envelope Sequence of enFeLV: Clues to the Targets of the Feline Immune Response

It was noted some years ago that cats infected with FeLV-A frequently produce antibodies which specifically neutralize FeLV-C/Sarma, despite the low isolation rate for subgroup C viruses. Also, cats with lymphoid tumours lacking detectable exogenous FeLV expression had anti-FeLV-C neutralizing antibodies. Several possible explanations for these phenomena were advanced, including the possibility that FeLV-C viruses arose by recombination with the then poorly characterized enFeLV sequences and that enFeLV might be expressed in some cryptic form in FeLV-negative tumours (ONIONS et al. 1982). However, we have since learned that FeLV-C viruses can arise from FeLV-A by minor changes which are probably mutational rather than recombinational (see Sect. 2.1). Also, FeLV-C isolates are known to be serotypically diverse and two of them behaved almost identically to FeLV-A/Glasgow-1 in their reactivity with feline sera (RUSSELL and JARRETT 1987a). As indicated in Sect. 2.1, the key to this problem may be that FeLV-C/Sarma differs from the other subgroup C viruses in that its *env* gene-coding sequence resembles enFeLV rather than FeLV-A/Glasgow-1 in the antigenic C-terminal portion of gp 70 *env*. Neutralizing antibodies specific for any of the enFeLV-related epitopes would have registered as "FeLV-C-specific" in serological analyses which used FeLV-C/Sarma as the target (RUSSELL and JARRETT 1978b).

Another puzzling phenomenon which may prove comprehensible from recent information on the sequence and expression pattern of FeLV *env* genes is an apparently tumour-specific antigen which was detected on the surface of feline lymphosarcoma cells (feline oncornavirus-associated cell membrane antigen; FOCMA). The most recently published study concluded that there is a cell surface protein (p70) related to, but structurally distinct from, FeLV-C/Sarma gp 70 which is expressed on the surface of all feline lymphoid tumours, regardless of the presence of exogenous FeLV (H.W. SNYDER et al. 1983). Expression of enFeLV *env* gene products on the tumour cells was proposed as the most likely explanation, although the mechanism by which this occurred was not addressed. Recently, we have found evidence that transcription of truncated enFeLV loci is a consistent feature of feline lymphoid tumours, including two cell lines lacking detectable exogenous FeLV sequences (J. NEIL and A. MCDOUGALL, unpublished results).

2.2 FeLV LTR Variation

In contrast to its murine viral relatives (QUINT et al. 1984), there is no evidence that FeLV LTR sequence variation arises by recombination with endogenous proviruses. However, specific changes to the FeLV LTR frequently accompany progression to leukaemia.

2.2.1 Enhancer Duplication

The majority of FeLV proviruses isolated from primary T-cell tumour DNA show tandem direct repeats of LTR enhancer sequences. This has been shown by DNA sequence analysis of molecularly cloned complete proviruses and by in situ amplification of the LTR enhancer domain using the PCR (FULTON et al. 1990). Notably, tumours induced by the highly leukaemogenic Rickard strain of FeLV contain enhancer duplicated viruses to the virtual exclusion of single-copy structures. The Rickard virus was obtained by sequential in vivo passage of thymic tumour extracts (ROJKO et al. 1979), which presumably selected for viruses with the duplicated enhancer structure. The duplications are precisely analogous to those which are found in murine leukaemia viruses with accelerated leukaemogenic potential (LI et al. 1987; CHATTOPADHYAY et al. 1989; HOLLAND et al. 1989).

The structure of the FeLV LTR is shown in outline in Fig. 2. The central enhancer motif is very well conserved among the mammalian C-type retroviruses (GOLEMIS et al. 1990). The basic motif is a tandem series of binding sites for nuclear factors LvB, core enhancer binding protein (SPECK and BALTIMORE 1987; or SEF-1; THORNELL et al. 1988), nuclear factor 1 (NF-1; SPECK and BALTIMORE 1987; NILSSON et al. 1989), and a glucocorticoid response element (MIKSICEK et al. 1986). At least in the LTR of FeLV T17T-31 which has been analysed at this level, the repeats appear to be duplications of discrete binding sites with no new sites generated at the junctions (FULTON et al. 1990). The termini of the repeats define each as a unique structure, except where the proviruses have been cloned from the same tumour (the T17 series) or cloned from separate tumours induced with the same viral inoculum (as in T8 and T10 induced by FeLV-Rickard).

Duplication of the core of the LTR enhancer might be expected to increase its potency. Consequently, by driving viral transcription to higher levels, viruses with duplicated enhancers would be expected to outgrow those without duplications. If the virus carries an oncogene, this would also be expressed at higher levels. However, functional assays of LTR activity using various reporter genes show only modest increases in activity resulting from the duplications (PLUMB et al. 1991). This is perhaps surprising in view of the strong evidence for disease acceleration in vivo (LI et al. 1984; CHATTOPADHYAY et al. 1989; FULTON et .al. 1990).

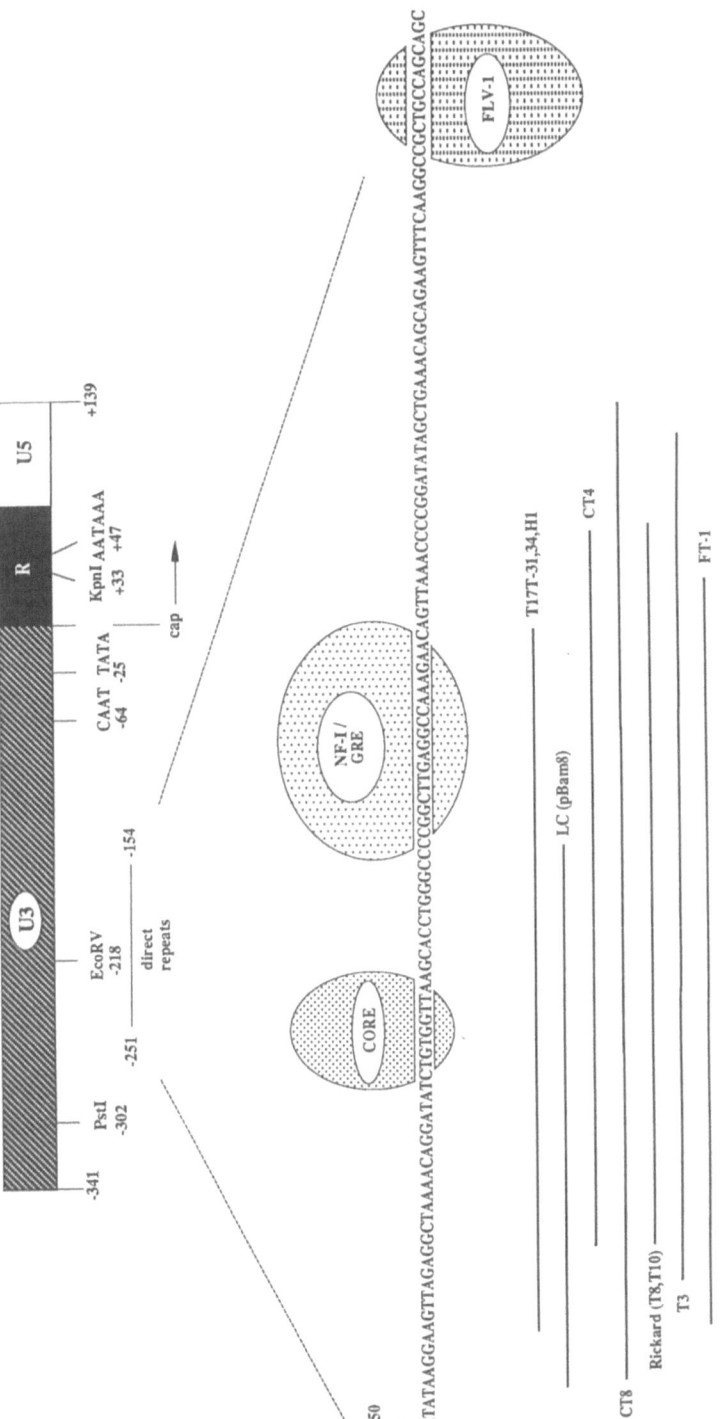

Fig. 2. Outline structure of the FeLV LTR with coordinates based on FeLV-A/Glasgow-1 (Stewart et al. 1986b). As discussed in the text, primary T-cell tumour isolates and highly leukaemogenic variants of FeLV frequently have duplications of the LTR enhancer core, including a cluster of highly conserved binding sites for feline nuclear proteins. Only the LvB/SV core cluster is present in all of the direct repeats (Fulton et al. 1990)

2.2.2 Enhancer Duplication and Insertional Mutagenesis

Although enhanced viral replication might account for disease acceleration mediated by the LTR duplications, another possibility is that by increasing enhancer potency, duplication augments *cis*-activation of cellular proto-oncogenes or increases the range over which it is effective. In support of this model, it appears that proviruses integrated upstream of feline c-*myc* in thymic lymphosarcomas may be almost exclusively enhancer duplicated structures. The only case examined directly by molecularly cloning showed a proviral LTR with three copies of the enhancer sequence (MIURA et al. 1989). Other proviruses at c-*myc* have not been analysed directly, but PCR analyses of tumour DNA indicate that tumours induced by FeLV-Rickard which have insertions at c-*myc* contain mainly enhancer-duplicated LTRs, suggesting that those at the c-*myc* locus will be of similar structure (FULTON et al. 1990; R. FULTON, unpublished results). Two notable exceptions in which PCR analysis showed only the unit length enhancer were T7, a field case tumour in which proviral integration was particularly close to the c-*myc* promoters and T24, the only known case of FeLV LTR promoter insertion at c-*myc* (FORREST et al. 1987). Whilst cellular gene activation may be rendered more efficient by enhancer duplication, it may not be required if proviral insertion is very close to the proto-oncogene promoter (as in T7) or accompanied by proviral deletions (as in T24).

Other tumour-specific FeLV integration loci have received little study so far, but one has been identified in a less common form of feline tumour, splenic lymphoma. Five out of seven tumours of this type had rearrangements at the same locus, which has been called *Flvi*-1 (L.S. LEVY, personal communication). The proviral structures at *Flvi*-1 have not yet been characterized in detail.

2.2.3 Other LTR Mutations

The in vitro transcriptional activity of FeLV LTRs shows significant heterogeneity, not all of which can be attributed to enhancer duplications. Few of the functional differences have as yet been ascribed to specific sequence changes, but it appears that two point mutational differences within the LTR enhancer domain of the FAIDS variant 1161C compared to 1161E confer increased replication in the 3201B T cell line (J. MULLINS, personal communication). Also, a cytopathic effect of FeLV-C/Sarma in 3201B cells correlates with increased viral transcription. In the same cell line, reporter gene constructs driven by the LTR of FeLV-C/Sarma are more active than those from the noncytopathic FeLV-A (ROJKO et al., submitted for publication). The LTRs of these two viruses differ by only a few point mutations (RIEDEL et al. 1986; STEWART et al. 1986b).

A well-characterized example of mutational loss of LTR activity is the LTR of proviral clone pT17T-22, which contains the transduced v-*tcr* gene (FULTON et al. 1987). This carries a mutation in the NF1 consensus binding site which abolishes in vitro binding and greatly reduces LTR activity. Intriguingly, activity is less impaired in tumour cell lines that have significantly lower levels of NF1 activity (PLUMB et al.).

2.3 FeLV Recombinants Carrying Host Cell Genes

2.3.1 FeLV Transduction of Cellular Genes: A Feature of Naturally Occurring Tumours

The best known aspect of FeLV oncogenesis is the frequency with which oncogene-containing variants have been isolated. A list of the host gene-containing isolates of FeLV is presented in Table 2. In view of the relatively meagre number of such cases which have emerged from other retroviral models (with the obvious exception of the avian leukosis/sarcoma viruses), it might be considered that FeLV is unusually promiscuous in recombination with cellular genes or that transduction by FeLV occurs in close conjunction with insertional mutagenesis (BISHOP 1987). However, FeLV recombination with host cell genes other than the closely related endogenous viral sequences has never been demonstrated in vitro, nor is there any experimental tumour system in which FeLV transduction of host cell genes has been observed. It would seem likely that the observed prevalence of transduction in FeLV tumours has been due to the selection for study of rare sporadic tumours from the natural population. Every single isolate listed in Table 2 was derived from a naturally occurring feline tumour. Where tested, each of the recombinant viruses carrying host cell genes has induced tumours similar to that from which it was isolated, that is fibrosarcomas (S.P. SNYDER and THEILEN 1969; GARDNER et al. 1970; McDONOUGH et al. 1971; BESMER 1983; BERGOLD et al. 1987) or T-lymphoid tumours (LEVY et al. 1988; ONIONS et al. 1987).

While there is no evidence to support the assertion that FeLV is an unusually recombinogenic retrovirus, there are apparent sequence preferences in the recombination points between virus and cellular sequence. These are discussed below.

Table 2. FeLV recombinants carrying host cell genes

Isolate	Host gene	Isolate	Host gene
a) From fibrosarcomas		b) From thymic lymphosarcomas	
FeSV-HZ2	abl	FeLV-LC	c-myc
FeSV-GA	fes	FeLV-FTT	c-myc
FeSV-ST	fes	FeLV-GT3	c-myc
FeSV-HZ1	fes	FeLV-GT11	c-myc
FeSV-GR	fgr, actin	FeLV-84929	c-myc
FeSV-TP1	fgr	FeLV-T17M	c-myc
FeSV-SM	fms	FeLV-T17T	tcr
FeSV-HZ5	fms		
FeSV-HZ4	kit		

2.3.2 Mechanism of Transduction: Remaining Problems

The process of oncogene transduction by retroviruses is sufficiently rare that it cannot be consistently reproduced under laboratory conditions; models to explain transduction are largely predictions based on the final structures of the transducing viruses. Since a virus found in a tumour may have undergone multiple rounds of replication prior to tumour development, and may have been passaged extensively in vitro prior to molecular analysis, we cannot be entirely confident that the structures are informative with respect to the primary recombinational events. Apparently intermediate stages in the transduction process are relatively rarely observed, with the possible exception of avian leukosis virus insertions at c-erb-B (NILSEN et al. 1985). Analysis of FeLV tumours has so far revealed no obvious transduction intermediate. Most examples of proviral integration are upstream of c-myc in the opposite transcriptional orientation. In one case (GT24), a truncated provirus was found to be integrated within c-myc intron 1, and S1 nuclease analysis of tumour RNA indicated that c-myc transcription initiated from the FeLV LTR rather than from the normal c-myc promoters (FORREST et al. 1987). Restriction mapping data was previously interpreted as indicative of a 3′-truncated provirus within intron 1. However, direct analysis of the integration site by "inverse PCR" cloning has shown that the integration was in exon 1 and that the FeLV 3′ LTR and not the 5′ LTR provided the promoter (H. TSUJIMOTO and J. NEIL, unpublished results).

Whatever transduction mechanism is invoked, the frequency with which oncogene-containing viruses arise is the multiplicative product of several sequential recombinations of low probability. The model proposed for the genesis of Rous sarcoma virus (RSV) (SWANSTROM et al. 1983) is one of several possible variations on the themes of proviral integration, DNA deletion, co-packaging and illegitimate recombination (GOFF et al. 1980). An alternative mechanism to that proposed for RSV appears to be operative for Rous associated virus-1 (RAV-1) induced erythroblastosis in line 15_1 chickens, where read-through transcription and aberrant splicing into the c-erb-B sequence is commonly seen as the mode of insertional activation. From this configuration, recombination to yield an erb-B transducing virus appears relatively straightforward (NILSEN et al. 1985). Such a mechanism would seem less likely to lead to transduction by FeLV, in which the 5′ splice donor site is upstream of the presumptive RNA packaging signal so that readthrough transcripts would not be copackaged (FICHT et al. 1984; HAMPE et al. 1983).

The final step in the model for generation of RSV was proposed to be a template switch or illegitimate recombination involving reverse transcriptase (SWANSTROM et al. 1983). This proposal was supported by a study which showed that transforming virus could be regenerated efficiently after transfection of Ha-MSV proviruses lacking a 3′ LTR and rescue with helper MuLV (GOLDFARB and WEINBERG 1981). A more recent study suggests that the high rate of virus regeneration may have been an artefact of the DNA transfection procedure (GOODRICH and DUESBERG 1988). If template switching is the final transduction step, it presumably occurs at a lower frequency than previously considered. However, further support for the template switching model comes from the observation that the PRCII avian sarcoma virus v-*fps* gene contains a polyA sequence at the 3' virus-oncogene junction (HUANG et al. 1986).

2.3.3 Recombinational Junctions in FeLV

Inspection of the FeLV-c-*onc* recombinational junctions (Fig. 3) shows immediately that there is marked clustering, particularly with respect to the 5' junctions. While this may reflect the existence of recombinational "hot-spots" within FeLV, it is conceivable that constraints on the efficient expression of transduced cellular genes determine the clustering. The cluster of junctions at the *pol-env* junction, for example, occurs between a consensus splice acceptor

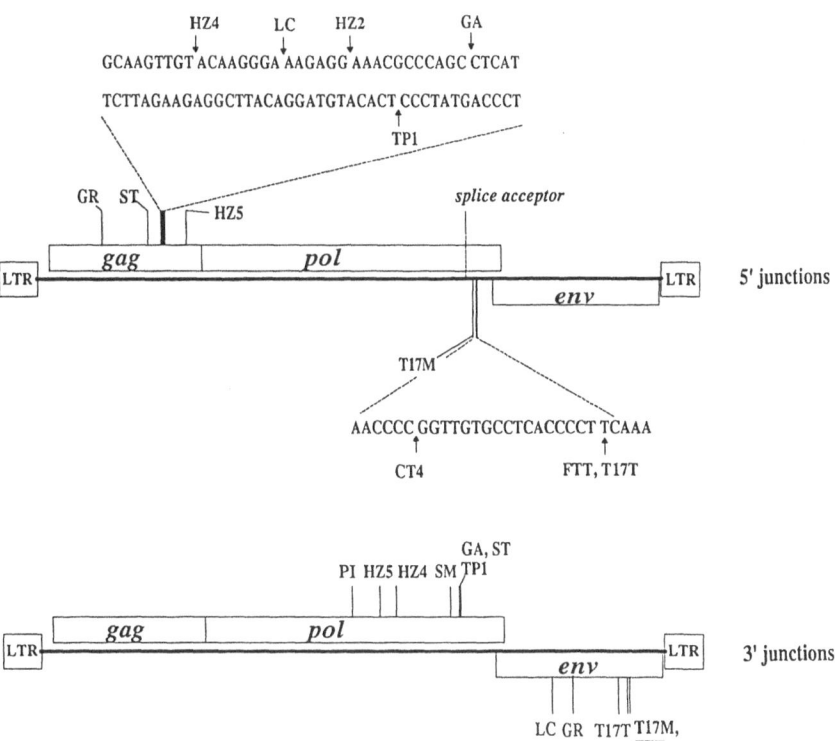

Fig. 3. Recombinational junctions in oncogene-containing feline retroviruses. The diagram is based on the published sequences of FeSV-GA (Gardner-Arnstein; HAMPE et al. 1982), FeSV- ST (Snyder-Theilen; HAMPE et al. 1982), FeSV-GR (Gardner-Rasheed; NAHARRO et al. 1984), FeSV-TP1 (KAPPES et al. 1989) FeSV-HZ (Hardy-Zuckerman) 2 (BESMER et al. 1986), HZ4 (BESMER et al. 1986), HZ5 (BESMER et al. 1987), FeLV- *myc* isolates LC (BRAUN et al. 1985), CT4 (STEWART et al. 1986a) FTT (DOGGETT et al. 1989), T17M (R. FULTON and J.C. NEIL, unpublished) and the T-cell receptor β-chain-containing recombinant T17T (FULTON et al. 1987)

and the initiation codon for the *env* gene product. This insertion site would be a prime location for expression of a heterologous gene product from its own initiation codon, and the structures of recombinants at that site are consistent with this mode of expression (DOGGETT et al. 1989; FULTON et al. 1987; STEWART et al. 1986a). Clusters of *gag* junction points could arise from constraints on the stability and cellular localization properties of *gag–onc* fusion proteins which are generated by recombination. Notwithstanding these cautions, some features of the nucleotide sequences at the junction points are of potential interest with regard to the recombination process (Fig. 4). In some cases low-fidelity matches to recombinase recognition sequences have been noted (DOGGETT et al. 1989; STEWART et al. 1986a). Fusion of proviral and proto-oncogene sequences mediated by DNA deletions may be favoured in cells with an active recombinase system, such as the developing cells of the immune

```
 5'    AGCAGAAC ACCCACCCC CGGGAGC       c-fes ex8
         I       IIIIIIIII IIIIIII
       TCCCCTTG ACCCACCCC CGGGAGC       ST-FeSV            (37, 94)
       IIIIIIII IIII IIII        I
       TCCCCTTG ACCCGCCCC AACTGGG       FeLV gag


       AGCCCCGG AGCAGC TGCCCGCGCG       c-fes ex1
         I      IIIIII IIIIIIIIII
       GAAACGCC AGCAGC TGCCCGCGCG       GA-FeSV            (37, 94)
       IIIIIIII IIIIII       I  I I
       GAAACGCC AGCAGC ATTTCTAGAG       FeLV gag


 3'    GATCGTAC CTCCTCC CCAGTCCAG       c-fes ex19
       IIIIIIII IIIIIII        I  I
       GATCGTAC CTCCTCC TGCCCCTGG       GA/ST-FeSV         (37, 94)
        I I      IIIIIII IIIIIIIII
       ATTGGGTG CTCCTCC TGCCCCTGG       FeLV pol


       TGTAGCAA CTCCTCAT ATCTGAAC       c-myc ex3
       IIIIIIII IIIIIIII        II
       TGTAGCAA CTCCTCAT GGGACTAC       FeLV-LC           (16, 109, 110)
        II   IIIIIIIII IIIIIIII
       AAGCCCAC CTCCTCAT GGGACTAC       FeLV env


       CACCTATT AGAGGGAGG AACTGGA       c-myc ex3
       IIIIIIII IIIIIIIII       I
       CACCTATT AGAGGGAGG GCTCTGT       FeLV-FTT, T17M     (21, 24, 109)
        I II   IIIIIIIIII IIIIIII
       TTCTTACA AGAGGGAGG GCTCTGT       FeLV env
```

Fig. 4. Virus-host sequence homologies at FeLV–oncogene junctions

system. However, most retrovirus–oncogene junctions do not conform to the recognition sequence.

In two cases, short stretches of homology are seen between FeLV and the c-*onc* sequence at the recombinational junction (ST and GA) (ROEBROCK et al. 1987). One of these (GA) is within a cluster of five recombinational junctions. However, most of the 5' junctions show no clear homology between virus and c-*onc*.

The sequences around the two clusters of 5' junctions are shown in detail in Fig. 3. Two motifs which can be found widely at retrovirus–oncogene junctions have been pointed out in these sequences. These motifs (GAGG/CCTC, BESMER et al. 1986; ACCCC, DOGGETT et al. 1989) occur with unusual frequency at *gag-onc* junctions, suggesting that they may act as target sites for some enzyme system which promotes recombination. Both motifs represent and occur within regions of pyrimidine/purine strand bias. It is interesting to note that strand bias is also a feature of the LTR terminal inverted repeat (TGAAAGACCCCC) which is probably recognized by the virus-specific enzyme that mediates the integration process. Similar sequence features have been identified close to the junctions of sequence duplications in the LTR enhancer (FULTON et al. 1990). Aberrant integrations or internal proviral cleavage might be promoted by genomic sequences resembling the terminal repeats. It is worth noting, however, that short perfect repeats are scattered throughout the FeLV *gag* gene and that these include a central motif homologous to the LTR inverted repeat (LAPREVOTTE 1989).

The recorded 3' FeLV-*onc* junctions show less marked clustering than the 5' junctions, although perhaps not surprisingly they are all within the 3' half of the genome. Precise coincidences of the 3' junction in both FeLV and the c-*onc* gene are seen in two isolates containing the c-*fes* sequence (GA and ST; ROEBROCK et al. 1987) and in two independent transductions of c-*myc* (FTT and T17M; DOGGETT et al. 1989; R. FULTON and J. NEIL, unpublished results). Both of these junction sites represent extended regions of homology between virus and c-*onc* sequence (seven and nine base pairs, respectively) and demonstrate pyrimidine/purine strand bias. Another junction showing homology between FeLV and c-*onc* is seen in the LC v-*myc* isolate, where again strand bias is a prominent feature. Both this and the T17M–FTT junction are notable for the occurrence of the GAGG/CCTC and ACCCC motifs within and adjacent to the recombination site. However, based on the transduction models already referred to (SWANSTROM et al. 1983), the 3' junctions may be a product of "forced recombination" after a hybrid virus-*onc* transcript has already been copackaged with helper viral RNA and may occur preferentially at sites of homology and/or strand bias.

2.3.4 Recombination Sites in c-*myc*

Figure 5 shows a diagram of the feline c-*myc* locus, illustrating proviral insertion sites and virus/*onc* junctions in FeLV/v-*myc* recombinant viruses. Nine proviral

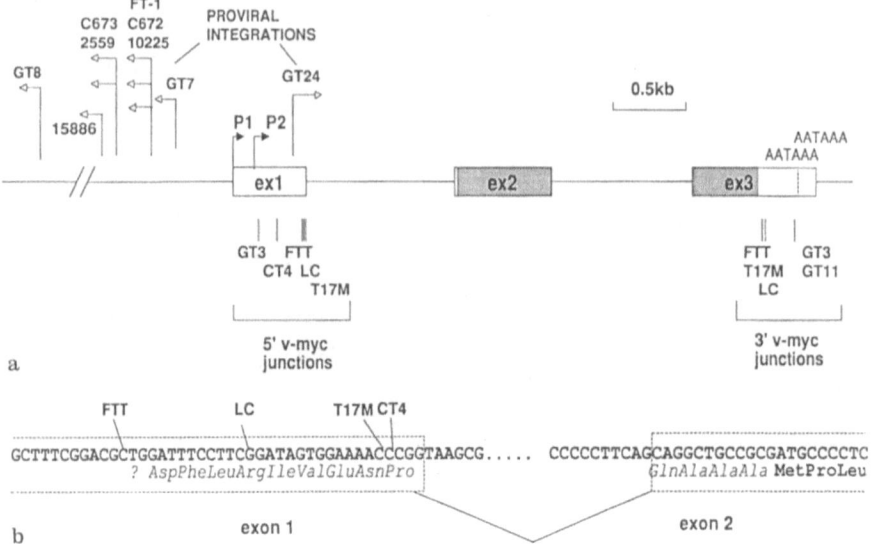

Fig. 5. a Proviral insertions at the feline c-*myc* gene and recombinational junctions in recombinant FeLV isolates carrying transduced c-*myc* genes. Sites of proviral insertion have been characterized in several studies (NEIL et al. 1987; FORREST et al. 1987; MIURA et al. 1989). The c-*myc* recombinational junction points are based on published analyses of the sequence and structure of the feline c-*myc* locus (STEWART et al. 1986a), sequences of FeLV recombinants carrying v-*myc* genes; CT4 (STEWART et al. 1986a), LC (BRAUN et al. 1985), FTT (DOGGETT et al. 1989), and T17M (R. FULTON and J.C. NEIL, unpublished results). Additional junction points have been mapped by S1 nuclease analysis of tumour cell RNA (FORREST et al. 1987). The coding sequence for the major c-*myc* gene product is represented by *hatched boxes* within exons 2 and 3. **b** Corrected sequence at the exon 1/exon 2 junction of feline c-*myc* showing recombinational junctions for FeLV-*myc* isolates (as in **a**) and a possible open reading frame extending from exon 1 as discussed in the text

insertions at c-*myc* have been mapped in the course of several independent studies (FORREST et al. 1987; NEIL et al. 1987; MIURA et al. 1989). In addition, four v-*myc*-containing FeLVs have been analysed at the primary sequence level (LC, CT4, FTT, T17M) (BRAUN et al. 1985; DOGGETT et al. 1989; STEWART et al. 1986a; R. FULTON and J. NEIL, unpublished) and further v-*myc* recombinational junctions have been mapped within c-*myc* by S1 nuclease analysis of tumour cell RNA (GT3, GT11) (FORREST et al. 1987).

As discussed already for the recombinational sites in FeLV, the FeLV–c-*myc* junctions are clustered to some extent. However, significant constraints apply to oncogenic activation of the mammalian c-*myc* genes. Truncation of the *myc* coding sequence at either end impairs or abolishes oncogenic activity (ENRIETTO 1989; HEANEY et al. 1986; SARID et al. 1987; STONE et al. 1987), providing a clear rationale for the occurrence of FeLV–*myc* junctions outside the coding domain.

A correction to our previously published sequence data is made here (STEWART et al. 1986a). The mammalian c-*myc* genes appear to encode two forms of the c-*myc* protein product which differ at the N terminus (HANN et al.

1987). The longer c-*myc* reading frame extends into exon 1 where a non-ATG codon defines the putative initiation site. Figure 5 shows the exon 1 and exon 2 sequences of feline c-*myc* and the relative positions of FeLV–*myc* 5′ recombinational junctions. In contrast to our published sequence, which suffered two errors in compilation, there is an open reading frame extending from a putative upstream translational start in exon 1 and in this respect the feline gene resembles the other mammalian c-*myc* genes. Of the v-*myc* sequences available for comparison, only FTT retains sufficient exon 1 sequences to encode this product and in that case the replacement of sequences 5′ to the initiation site with FeLV sequences may affect recognition of the non-AUG codon which is thought to be utilized (HANN et al. 1987).

The exon 1 content of feline v-*myc* genes is variable and is most extensive in the GT3 isolate, which has not been molecularly cloned but which has been demonstrated by S1 nuclease analysis of RNA from tumour cells to contain a contiguous 95-base fragment of exon 1 extending from an exon 1 *Sma*l site almost to the p2 promoter (FORREST et al. 1987). The GT3 isolate encodes a *gag-myc* readthrough product (J. NEIL, unpublished results) which must include the exon 1 sequences unless these are spliced out. The presence in GT3 v-*myc* of c-*myc* exon 1 sequences from 3′ to the *Sma*l site has not been addressed, but other v-*myc* genes which have been sequenced directly are deficient in 3′ exon 1 sequences, including the CT4 v-*myc* gene in which an internal deletion within exon 1 appears to have occurred (STEWART et al. 1986a). A possible constraint on the presence of 3′ exon 1 sequences in v-*myc* is the transcriptional elongation block which appears to operate at the homologous site in human c-*myc* (BENTLEY and GROUDINE 1988).

Within the exon 3 untranslated sequences, all of the 3′ FeLV-*myc* junctions precede the major c-*myc* polyadenylation signal (FORREST et al. 1987).

In conclusion, clustering of recombination sites within c-*myc* seems more likely to result from constraints on the structure and function of v-*myc* genes than from the operation of an efficient recombinational mechanism. The probable constraints are the requirement for the entire *myc* coding sequence to generate a functional product and perhaps the inhibitory effect of transcriptional or translational controls conferred by c-*myc* exon 1.

3 Conclusions

3.1 Lessons from a Natural Leukaemogenesis Model: Comparison of the Feline and Murine C-Type Retroviruses

Numerous aspects of this review extend the parallel between FeLV and the murine C-type retroviruses. However, contrasts between the two systems may be instructive.

Many studies on MuLV leukaemogenesis have used the AKR mouse system, in which it appears that an ordered set of changes in the viral *env* gene and the LTR are obligatory for tumour development (Teich et al. 1982). We suspect that this situation is largely an artefact of the inbreeding which led to the isolation of this high leukaemia incidence mouse strain. The AKR mouse contains in its genome a replication-competent, ecotropic virus (Akv) which is expressed early in development (Rowe and Pincus 1972). From this context the AKR virus must overcome several handicaps before it becomes leukaemogenic. The AKR virus may present a barrier to its own replication by virtue of viral interference mediated either by complete virus expression or the expression of *env* gene products. Also, mice are not immunologically tolerant to the endogenous virus (Huebner et al. 1976). The generation of MCF recombinant viruses, which use a different receptor on mouse cells (Rein 1982), presumably overcomes this barrier and thereby aids virus spread and further recombination. The requirement for MCF virus formation is indicated by the fact that AKR leukaemias contain no new ecotropic viral integrations (Yoshimura and Breda 1981). The second handicap is that the germline AKR virus appears to be disabled by a mutation in the enhancer core which restricts its expression, particularly in T cells (Boral et al. 1989). Repair of this mutation presumably occurs during preleukaemic virus replication in the mouse, most likely by recombination with a xenotropic endogenous provirus which donates the MCF LTR (Quint et al. 1984).

As a successful horizontally transmitted virus, FeLV shares few of the handicaps discussed for Akv. The commonly isolated FeLV strains have enhancer core sequences identical to those of leukaemogenic MuLV which function efficiently in T cells, and the exogenous FeLV replicates in a relatively unrestricted fashion. While the generation of *env* gene or LTR mutants of FeLV may augment the pathogenicity of FeLV (see Sect. 2.1, 2.2), neither appears to be obligatory. From our studies, naturally occurring FeLV-positive tumours invariably contain the ecotropic FeLV-A *env* sequence, while only a portion have detectable new integrations by the *env* gene recombinant FeLV-B viruses (J. Neil, unpublished).

A second contrast between FeLV and MuLV oncogenesis is in the recorded frequency of transduction versus insertional activation of cellular oncogenes. However, the Rickard strain of FeLV behaves similarly to several MuLV isolates which demonstrate exclusively the insertional activation mode at the c-*myc* locus and no examples of transduction. As discussed in Sect. 2.2, the LTR enhancer duplication of FeLV-Rickard most likely promotes efficient insertional activation. Most of the MuLV experiments in which c-*myc* insertion has been recorded employed viruses with similar enhancer duplications (Corcoran et al. 1984; Selten et al. 1984; Li et al. 1987) and would therefore be expected to generate a similar bias towards enhancer insertion as the primary oncogenic mechanism. A telling observation in this respect is that examination of sporadic tumours of wild mice has revealed more examples of oncogene transduction (Langon et al. 1989), (H.C. Morse, personal communication).

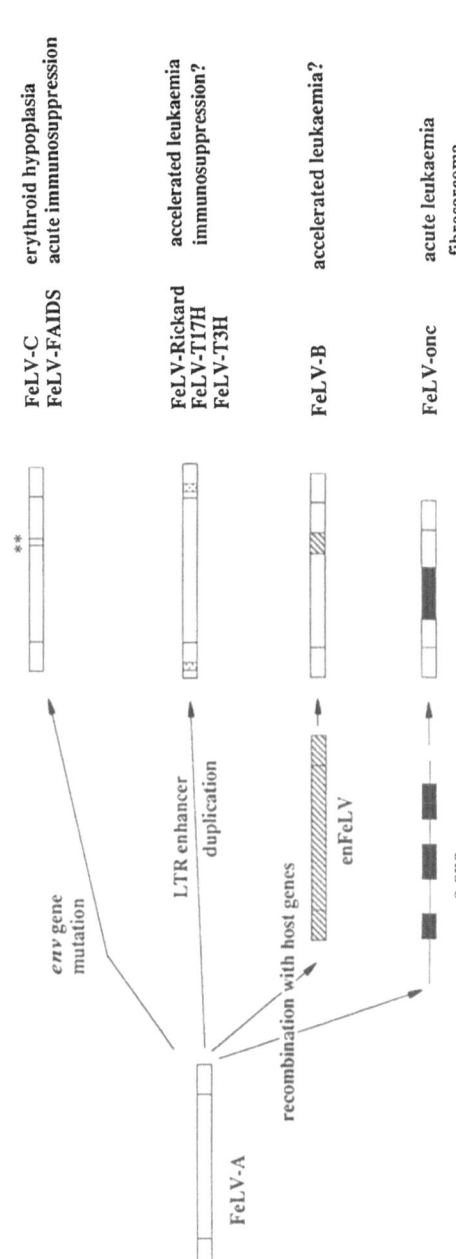

Fig. 6. The origins of FeLV variants and their associated diseases

3.2 Prospects

The study of feline immune responses to FeLV has declined in recent years. This is regrettable since our knowledge of FeLV molecular biology has only recently matured to the point where it can make a useful contribution. Some progress has been made in mapping neutralization epitopes in the FeLV *env* gene using murine monoclonal antibodies (NUNBERG et al. 1984a) and rabbit anti-peptide sera (ELDER et al. 1987), but we have no precise information on the feline immune response at this level. It will be of particular interest to monitor the response of cats to the enFeLV-related epitopes discussed in Sect. 2.2, since these may be expected to be expressed either as normal lymphoid cell antigens or as tumour neoantigens.

Another area where the unique biology of FeLV invites further study is the molecular characterization of host cell receptors for FeLV A, B and C. It may be expected that identification of these molecules will greatly aid our understanding of the establishment and progression of FeLV disease. The recent identification of a candidate receptor for ecotropic MuLV (ALBRITTON et al. 1989) gives further encouragement.

In conclusion, FeLV provides us with an example of a virus which successfully coexists with its host despite its considerable pathogenicity. In doing so it has evolved a balance between self-propagation and rapid destruction of the host's immune competence; highly immunosuppressive or oncogenic variants probably die out with their host. The known variants are summarized in Fig. 6. Whilst variant generation appears to underline much of the acute pathogenesis of FeLV, it must be recognized that any one variant represents only a narrow section across the broad range of possible interactions between virus and host. For a wider view, the comparative study of retroviral pathogenesis in natural populations is required.

References

Abkowitz JL, Holly RD, Adamson JW (1987a) Retrovirus-induced feline pure red cell aplasia: the kinetics of erythroid marrow failure. J Cell Physiol 132: 571–577

Abkowitz JL, Holly RD, Grant CK (1987b) Retrovirus-induced feline pure red cell aplasia. Hematopoietic progenitors are infected with Feline leukemia virus and erythroid burst-forming cells are uniquely sensitive to heterologous complement. J Clin Invest 80: 1056–1063

Albritton LM, Tseng L, Scadden D, Cunningham JM (1989) A putative murine ecotropic retrovirus receptor gene encodes a multiple membrane-spanning protein and confers susceptibility to virus infection. Cell 57: 659–666

Bentley DL, Groudine M (1988) Sequence requirements for premature termination of transcription in the human c-*myc* gene. Cell 53: 245–256

Benveniste RE, Todaro GJ (1975) Segregation of RD-114 FeLV-related sequences in crosses between domestic cat and leopard cat. Nature 257: 506–508

Benveniste RE, Sherr CJ, Todaro GJ (1975) Evolution of type C viral genes: origin of feline leukemia virus. Science 190: 886–888

Bergold PJ, Blumenthal JA, Dandrea E, Snyder HW, Lederman L, Silverstone A, Nguyen H, Besmer P (1987) Nucleic acid sequence and oncogenic properties of the HZ2 feline sarcoma virus v-*abl* insert. J Virol 61: 1193–1202

Berry BT, Ghosh AK, Kumar DV, Spodick DA, Roy-Burman P (1988) Structure and function of endogenous feline leukemia virus long terminal repeats and adjoining regions. J Virol 62: 3631–3641

Besmer P (1983) Acute transforming feline retroviruses. Curr Top Microbiol Immunol 107: 1–27

Besmer P, Hardy WD Jr, Zuckerman EE, Bergold P, Lederman L, Snyder HW Jr (1983a) The Hardy-Zuckerman 2-FeSV, a new feline retrovirus with oncogene homology to Abelson-MULV. Nature 303: 825–828

Besmer P, Snyder HW Jr, Murphy JE, Hardy WD Jr, Parodi A (1983b) The Parodi-Irgens feline sarcoma virus and simian sarcoma virus have homologous oncogenes, but in different contexts of the viral genomes. J Virol 46: 606–613

Besmer P, Lader E, George PC, Bergold PJ, Qiu FH, Zuckerman EE, Hardy WD (1987) A new acute transforming feline retrovirus with *fms* homology specifies a C-terminally truncated version of the c-*fms* protein that is different from SM-feline sarcoma virus v-*fms* protein. J Virol 60: 194–203

Besmer P, Murphy JE, George PC, Qui F, Bergold PJ, Lederman L, Snyder HW Jr, Brodeur D, Zuckerman EE, Hardy WD (1986) A new acute transforming feline retrovirus and relationship of its oncogene v-*kit* with the protein kinase gene family. Nature 320: 415–421

Bishop JM (1987) The molecular genetics of cancer. Science 235: 305–311

Boral AL, Okenquist SA, Lenz J (1989) Identification of the SL3-3 virus enhancer core as a T-lymphoma cell-specific element. J Virol 63: 76–84

Braun MJ, Deininger PL, Casey JW (1985) Nucleotide sequence of a transduced *myc* gene from a defective feline leukemia provirus. J Virol 55: 177–183

Busch MP, Devi BG, Soe LH, Perbal B, Baluda MA, Roy-Burman P (1983) Characterization of the expression of cellular retrovirus genes and oncogenes in feline cells. Hematol Oncol 1: 61–75

Casey JW, Roach A, Mullins JI, Burck KB, Nicolson MO, Gardner MB, Davidson N (1981) The U3 portion of feline leukemia virus DNA identifies horizontally acquired proviruses in leukemic cats. Proc Natl Acad Sci USA 77: 7778–7782

Chattopadhyay SK, Baroudy BM, Holmes KL, Frederickson TN, Lander ML, Morse HC III, Hartley JW (1989) Biologic and molecular genetic characteristics of a unique MCF virus that is highly leukemogenic in ecotropic virus-negative mice. Virology 168: 90–100

Corcoran LM, Adams JM, Dunn AR, Cory S (1984) Murine T lymphomas in which the cellular *myc* oncogene has been activated by retroviral insertion. Cell 37: 113–122

Doggett DL, Drake AL, Hirsch V, Rowe ME, Stallard V, Mullins JI (1989) Structure, origin and transforming activity of the feline leukemia virus- *myc* recombinant provirus, FTT. J Virol 630: 2108–2117

Donahue PR, Hoover EA, Beltz GA, Riedel N, Hirsch V, Overbaugh J, Mullins JI (1988) Strong sequence conservation among horizontally transmissible, minimally pathogenic feline leukemia viruses. J Virol 62: 722–731

Elder, JH, Mullins JI (1983) Nucleotide sequence of the envelope gene of Gardner-Arnstein feline leukemia virus B reveals unique sequence homologies with a murine mink cell focus-forming virus. J Virol 46: 871–880

Elder JH, Mullins JI (1985) Nucleotide sequence of the envelope genes and LTR of the subgroup B, Rickard strain of feline leukemia virus. In: Weiss R, Teich N, Varmus H, Coffin J (eds) RNA tumor viruses. Cold Spring Harbor Laboratory, New York, pp 1005–1010

Elder JH, McGee JS, Munson M, Houghten RA, Kloetzer W, Bittle JL, Grant CK, (1987) Localization of neutralizing regions of the envelope gene of Feline leukemia virus using anti-synthetic peptide antibodies. J Virol 61(1): 8–15

Enrietto PJ (1989) A small deletion in the carboxy terminus of the viral *myc* gene renders the virus partially transformation defective in avian fibroblasts. Virology 168: 256–266

Ficht T, Chang L-J, Stolzfus (1984) Avian sarcoma virus *gag* and *env* gene structural protein precursors contain a common amino terminus. Proc Natl Acad Sci USA 81: 362–366

Forrest D, Onions D, Lees G, Neil JC (1987) Altered structure and expression of c-*myc* in feline T-cell tumours. Virology 158: 194–205

Franchini G, Gelmann EP, Dalla-Favera R, Gallo RC, Wong-Staal F (1982) Human gene (c-*fes*) related to the onc sequences of Snyder-Theilen feline sarcoma virus. Mol Cell Biol 2: 1014–1019

Fulton R, Forrest D, McFarlane R, Onions D, Neil JC (1987) Retroviral transduction of T-cell antigen receptor *B-chain* and *myc* genes. Nature 326: 190–194

Fulton R, Plumb M, Shield L, Neil JC (1990) Structural diversity and nuclear protein binding sites in the long terminal repeats of feline leukaemia virus. J Virol 64: 1675–1682

Gardner MB, Rongey RW, Arnstein P, Estes JD, Sarma P, Huebner RJ, Rickard CG (1970) Experimental transmission of feline fibrosarcoma to cats and dogs. Nature 226: 807–809

Goff SP, Gilboa E, Witte ON, Baltimore D (1980) Structure of the Abelson murine leukemia virus genome and the homologous cellular gene: studies with cloned viral DNA. Cell 22: 777–785

Goldfarb MP, Weinberg RA (1981) Generation of novel, biologically active Harvey sarcoma viruses via apparent illegitimate recombination. J Virol 38: 136–150

Golemis EA, Speck NA, Hopkins N (1980) Alignment of U3 region sequences of mammalian type C viruses: identification of highly conserved motifs and implications for enhancer design. J Virol 64: 534–542

Goodrich DW, Duesberg PH (1988) Retroviral transduction of oncogenic sequences involves viral DNA instead of RNA. Proc Natl Acad Sci USA 85: 3733–3737

Hampe A, Laprevotte I, Galibert F (1982) Nucleotide sequences of feline retroviral oncogenes (v-*fes*) provide evidence for a family of tyrosine-specific protein kinase genes. Cell 30: 775–785

Hampe A, Gobet M, Even J, Sherr CJ, Galibert F (1983) Nucleotide sequences of feline sarcoma virus long terminal repeats and 5' leaders show extensive homology to those of other mammalian retroviruses. J Virol 45: 466–472

Hampe A, Gobet M, Sherr CJ, Galibert F (1984) Nucleotide sequence of the feline retroviral oncogene v-*fms* shows unexpected homology with oncogenes encoding tyrosine-specific protein kinases. Proc Natl Acad Sci USA 81: 85–89

Hann SR, King MW, Bentley DL, Anderson CW, Eisenman RN (1987) A non-AUG translational initiation in c-*myc* exon 1 generates an N-terminally distinct protein whose synthesis is disrupted in Burkitt's lymphomas. Cell 52: 185–195

Hardy WD Jr, Hess PW, MacEwen EG, McClelland AJ, Zuckerman EE, Essex M, Cotter SM, Jarrett O (1976) Biology of feline leukemia virus in the natural environment. Cancer Res 36: 582–588

Hardy WD Jr (1980) Feline leukemia virus diseases. In: Hardy WD Jr, Essex M, McCelland AJ (eds) Feline leukemia virus. Elsevier/North Holland, New York, pp 3–31

Heaney ML, Pierce J, Parsons JT (1986) Site-directed mutagenesis of avian myelocytomatosis virus 29: biological activity and intracellular localization of structurally altered proteins. J Virol 60: 167–176

Holland CA, Thomas CY, Chattopadhyay SK, Koehne C, O'Donnell PV (1989) Influence of enhancer sequences on thymotropism and leukemogenicity of mink cell focus-forming viruses. J Virol 63: 1284–1292

Hoover EA, Kociba GJ, Hardy WD Jr, Yohn DS (1974) Erythroid hypoplasia in cats inoculated with feline leukemia virus. J Natl Cancer Inst 53(5): 1271–1276

Huang CC, Hay N, Bishop JM (1986) The role of RNA molecules in transduction of the proto-oncogene c-*fps*. Cell 44: 935–940

Huebner RJ, Gilden RV, Toni R, Hill RW, Trimmer RW, Fish DC, Sass B (1976) Prevention of spontaneous leukemia in AKR mice by type-specific immunosuppression of endogenous ecotropic virogenes. Proc Natl Acad Sci USA 73: 4633–4635

Jarrett O (1984) Pathogenesis of feline leukaemia virus-related diseases. In: Goldman JM, Jarrett O (eds) Mechanisms of viral leukaemogenesis. Churchill-Livingstone, Edinburgh, pp 135–154

Jarrett O, Russell PH (1978) Differential growth and transmission in cats of feline leukaemia viruses of subgroups A and B. Int J Cancer 21: 466–472

Jarrett O, Hardy WDJr, Golder MC, Hay D (1978a) The frequency of occurrence of feline leukemia virus subgroups in cats. Int J Cancer 21: 334–337

Jarrett O, Laird HM, Hay D (1973) Determinants of the host range of feline leukaemia viruses. J Gen Virol 20: 169–175

Jarrett O, Russell PH, Hardy WD Jr (1978b) The influence of virus subgroup on the epidemiology of feline leukaemia virus. In: Bentveltzen P (ed) Advances in comparative leukemia research. Elsevier, Amsterdam, pp 25–28

Jasin M, Berg P (1988) Homologous integration in mammalian cells without target gene selection. Genes Dev 2: 1353–1363

Kappes B, Ziemiecki A, Muller RG, Theilen GH, Bauer H, Daniekow A (1989) The TP1 isolate of feline sarcoma virus encodes a *fgr*-related oncogene lacking gamma actin sequences. Oncogene 4: 363–367

Kawakami TG, Theilen GH, Dungworth DL, Munn RJ, Beall SG (1967) C-type viral particles in plasma of cats with feline leukemia. Science 158: 1049

Khan AS (1984) Nucleotide sequence analysis establishes the role of endogenous murine leukemia virus DNA segments in formation of recombinant mink cell focus-forming murine leukemia viruses. J Virol 50(3): 864–871

Kumar DV, Berry BT, Roy-Burman P (1989) Nucleotide sequence and distinctive characteristics of the env gene of endogenous feline leukemia provirus. J Virol 63: 2379–2384

Langdon WY, Hartley JW, Klinken SP, Ruscetti SK, Morse HC III (1989) V-cbl, an oncogene from a dual-recombinant murine retrovirus that induces early B- lineage lymphomas. Proc Natl Acad Sci USA 86: 1168–1172

Laprevotte I (1989) Scrambled duplications in the feline leukemia virus gag gene: a putative pattern for molecular evolution. J Mol Evol 29: 135–148

Levy LS, Gardner MB, Casey JW (1984) Isolation of a feline leukaemia provirus containing the oncogene myc from a feline lymphosarcoma. Nature 308: 853–856

Levy LS, Fish RE, Baskin GB (1988) Tumorigenic potential of a myc-containing strain of feline leukemia virus in vivo in domestic cats. J Virol 62: 4770–4773

Li Y, Holland CA, Hartley JW, Hopkins N (1984) Viral integration near c-myc in 10–20% of MCF 247-induced AKR lymphomas. Proc Natl Acad Sci USA 81:6808–6811

Li Y, Golemis E, Hartley JW, Hopkins N (1987) Disease specificity of nondefective Friend and Moloney murine leukemia viruses is controlled by a small number of nucleotides. J Virol 61: 693–700

Luciw P, Parker D, Potter S, Najarian R (1985) Feline leukemia virus (FeLV), strains A/Glasgow-1 and C, env genes. In: Weiss R, Teich N, Varmus H, Coffin J (eds) RNA tumor viruses. Cold Spring Harbor Laoratory, New York, pp 1000–1004

Mackey L, Jarrett W, Jarrett O, Laird HM (1975) Anemia associated with feline leukemia virus infection in cats. J Natl Cancer Inst 54: 209–217

McDonough SK, Larsen S, Brodey RS, Stock ND, Hardy WD Jr (1971) A transmissible feline fibrosarcoma of viral origin. Cancer Res 31: 953–956

Miksicek R, Heber A, Schmid W, Danesch U, Posseckert G, Beato M, Schutz G (1986) Glucocorticoid responsiveness of the transcriptional enhancer of moloney murine sarcoma virus. Cell 46: 283–290

Miura T, Shibuya M, Tsujimoto H, Fukusawa M, Hayami M (1989) Molecular cloning of a feline leukemia provirus integrated adjacent to the c-myc gene in a feline T-cell leukemia cell line and the unique structure of its long terminal repeat. Virology 169: 458–461

Mullins JI, Brody DS, Binari RC Jr, Cotter SM (1984) Viral transduction of c-myc gene in naturally occurring feline leukaemias. Nature 308: 856–858

Mullins JI, Chen CS, Hoover EA (1986) Disease-specific and tissue-specific production of unintegrated feline leukaemia virus variant DNA in feline AIDS. Nature 319: 333–336

Naharro G, Robbins KC, Reddy EP (1984) Gene product of v-fgr onc: hybrid protein containing a protein of actin and a tyrosine-specific protein kinase. Science 223: 63–66

Neil, JC (1984) Molecular aspects of feline leukaemia viruses and their related diseases. In: Rigby PW, Wilkie NM (eds) Viruses and cancer. Cambridge University Press, Cambridge, pp 219–239

Neil JC, Hughes D, McFarlane R, Wilkie NM, Onions DE, Lees G, Jarrett O (1984) Transduction and rearrangement of the myc gene by feline leukaemia virus in naturally occurring T-cell leukaemias. Nature 308: 814–820

Neil JC, Forrest D, Doggett DL, Mullins JI (1987) The role of feline leukaemia virus in naturally-occurring leukaemias. Cancer Sur 6: 117–137

Nilsen T, Maroney PA, Goodwin R, Rottman RM, Crittenden LB, Raines MA, Kung H-J (1985) c-erbB activation in ALV-induced erythroblastosis: novel RNA processing and promoter insertion results in expression of an amino-truncated EGF receptor. Cell 41: 719–726

Nilsson P, Hallberg B, Thornell A, Grundstrom T (1989) Mutant analysis of protein interactions with a nuclear factor I binding site in the SL3-3 virus. Nucleic Acids Res 17: 4061–4075

Niman HL, Gardner MB, Stephenson JR, Roy-Burman P (1977a) Endogenous RD-114 virus genome expression in malignant tissues of domestic cats. J Virol 23: 578–586

Niman HL, Stephenson JR, Gardner MB, Roy-Burman P (1977b) RD-114 and feline leukaemia virus genome expression in natural lymphomas of domestic cats. Nature 266: 357–360

Niman HL, Akhavi M, Gardner MB, Stephenson JR, Roy-Burman P (1980) Differential expression of two distinct endogenous retrovirus genomes in developing tissues of the domestic cat. J Natl Cancer Inst 64: 587–594

Nunberg JH, Rogers G, Gilbert JH, Snead RM (1984a) Method to map antigenic determinants recognized by monoclonal antibodies: localization of a determinant of virus neutralization on the feline leukemia virus envelope protein gp 70. Proc Natl Acad Sci USA 81: 3675–3679

Nunberg JH, Williams ME, Innis MA (1984b) Nucleotide sequences of the envelope genes of two isolates of feline leukemia virus subgroup B. J Virol 49: 629–632

Okabe H, Twiddy E, Gilden RV, Hatanaka M, Hoover EA, Olsen RG (1976) FeLV- related sequences in DNA from a FeLV-free cat colony. Virology 69: 798–801

Okabe H, DuBuy J, Gilden RV, Gardner MB (1978) A portion of the feline leukaemia virus genome is not endogenous in cat cells. Int J Cancer 22: 70–78

Onions D, Jarrett O, Testa N, Frassoni F, Toth S (1982) Selective effect of feline leukaemia virus on early erythroid precursors. Nature 296: 156–158

Onions D, Lees G, Forrest D, Neil J (1987) Recombinant feline viruses containing the *myc* gene rapidly produce clonal tumours expressing T-cell antigen receptor gene transcripts. Int J Cancer 40: 40–45

Overbaugh J, Donahue PR, Quackenbush SL, Hoover EA, Mullins JI (1988a) Molecular clonal of a feline leukemia virus that induces fatal immunodeficiency disease in cats. Science 239: 906–910

Overbaugh J, Riedel N, Hoover EA, Mullins JI (1988b) Transduction of endogenous envelope genes by feline leukaemia virus in vitro. Nature 332: 731–734

Plumb M, Fulton R, Braemer L, Willisen K, Neil JC (1991) Nuclear factor 1 activates the feline leukaemia virus long terminal repeat but is post-transcriptionally down-regulated in leukaemia cell lines. J Virol

Poss ML, Mullins JI, Hoover EA (1989) Post-translational modifications distinguish the envelope glycoprotein of the immunodeficiency disease inducing FeLV-FAIDS retrovirus. J Virol 63: 189–195

Quint W, Boelens W, van Wezenbeek P, Cuypers T, Robanus-Maandag E, Selten G, Berns A (1984) Generation of AKR mink cell focus-forming viruses: a conserved single-copy xenotrope-like provirus provides recombinant long terminal repeat sequences. J Virol 50: 432–438

Rein A (1982) Interference grouping of Murine leukemia viruses: a distinct receptor for the MCF-recombinant viruses in mouse cells. Virology 120: 251–257

Riedel N, Hoover EA, Gasper PW, Nicolson MO, Mullins JI (1986) Molecular analysis and pathogenesis of the feline aplastic anemia retrovirus, FeLV-C- Sarma. J Virol 60: 242–250

Riedel N, Hoover EA, Dornsife RE, Mullins JI (1988) Pathogenic and host range determinants of the feline aplastic anemia retrovirus. Proc Natl Acad Sci USA 85: 2758–2762

Roebroek AJM, Schalken JA, Onnekink C, Bloemers HPJ, Van de Ven WJM (1987) Structure of the Feline c-*fes*/*fps* proto-oncogene: genesis of a retroviral oncogene. J Virol 61: 2009–2116

Rojko JL, Hoover EA, Mathes LE, Olsen RG, Schaller JP (1979) Pathogenesis of experimental feline leukemia virus infection. J Natl Cancer Inst 63: 759–768

Rowe WP, Pincus T (1972) Quantitative studies of naturally-occurring murine leukemia virus infection of AKR mice. J Exp Med 135: 429–436

Russell PH, Jarrett O (1978b) The specificity of neutralizing antibodies to feline leukaemia viruses. Int J Cancer 21: 768–778

Russell PH, Jarrett O (1978a) The occurrence of feline leukaemia virus neutralizing antibodies in cats. Int J Cancer 22: 351–357

Sarid J, Halazonetis T, Murphy W, Leder P (1987) Evolutionarily conserved regions of the human c-*myc* protein can be uncoupled from transforming activity. Proc Natl Acad Sci USA 840: 170–173

Sarma PS, Log T (1973) Subgroup classification of feline leukemia and sarcoma viruses by viral interference and neutralization tests. Virology 54: 160–169

Sarma PS, Log T, Jain D, Hill PR, Huebner RJ (1975) Differential host range of viruses of feline leukemia-sarcoma complex. Virology 64: 438–446

Selten G, Cuypers HT, Zijlstra M, Melief C, Berns A (1984) Involvement of c-*myc* in MuLV-induced T-cell lymphomas in mice: frequency and mechanisms of activation. EMBO J 3: 3215–3222

Snyder HW Jr, Singhal MC, Zuckerman EE, Jones FR, Hardy WD Jr (1983) The feline oncornavirus-associated cell membrane antigen (FOCMA) is related to, but distinguishable from, FeLV-C gp70. Virology 131: 315–327

Snyder HW, Singhal MC, Zuckerman EE, Hardy WD (1984) Isolation of a new feline sarcoma virus (HZ1-FeSV): biochemical and immunological characterization of its translation product. Virology 132: 205–210

Snyder SP, Theilen GH (1969) Transmissible feline fibrosarcoma. Nature 221: 1074–1075

Soe LH, Devi BG, Mullins JI, Roy-Burman P (1983) Molecular cloning and characterization of endogenous feline leukemia virus sequences from a cat genomic library. J Virol 46: 829–840

Soe LH, Shimizu RW, Landolph JR, Roy-Burman P (1985) Molecular analysis of several classes of endogenous feline leukemia virus elements. J Virol 56: 701–710

Speck NA, Baltimore D (1987) Six distinct nuclear factors interact with the 75-base-pair repeat of the Moloney leukemia virus enhancer. Mol Cell Biol 7: 1101–1110

Stewart MA, Forrest D, McFarlane R, Onions D, Wilkie N, Neil JC (1986a) Conservation of the c-*myc* coding sequence in tranduced feline v-*myc* genes. Virology 154: 121–134

Stewart MA, Warnock M, Wheeler A, Wilkie N, Mullins JI, Onions DE, Neil JC (1986b) Nucleotide sequences of a feline leukemia virus subgroup A envelope gene and long terminal repeat and evidence for the recombinational orgin of subgroup B viruses. J Virol 58: 825–834

Stone J, Lange T, Ramsay G, Jakobovits E, Bishop JM, Varmus HE, Lee W (1987) Definition of regions in human c-*myc* that are involved in transformation and nuclear localization. Mol Cell Biol 7: 1697–1709

Swanstrom R, Parker RC, Varmus HE, Bishop JM (1983) Transduction of a cellular oncogene: the genesis of Rous sarcoma virus. Proc Natl Acad Sci USA 80: 2519–2523

Teich N, Wyke J, Mak T, Bernstein A, Hardy WD Jr (1982) Pathogenesis of retrovirus-induced disease. In: Weiss R, Teich N, Varmus H, Coffin J (eds) RNA tumor viruses. Cold Spring Harbor Laboratory, New York, pp 785–998

Testa NG, Onions D, Jarrett O, Frassoni F, Eliason JF (1983) Haemopoietic colony formation (BFU-E, GM-CFC) during the development of pure red cell hypoplasia induced in the cat by feline leukaemia virus. Leuk Res 7: 103–116

Thornell A, Halberg B, Grundstrom T (1988) Differential protein binding in lymphocytes to a sequence in the enhancer of the mouse retrovirus SL3-3. Mol Cell Biol 8: 1625–1637

Tzavaras T, Stewart M, McDougall A, Fulton R, Testa N, Onions DE, Neil JC (1990) Molecular cloning and characterisation of a defective recombinant feline leukaemia virus associated with myeloid leukaemia. J Gen Virol 71: 343–354

Yoshimura FK, Breda M (1981) Lack of AKR ecotropic provirus amplification in AKR leukaemic thymuses. J Virol 39: 808–815

Ziemiecki A, Hennig D, Gardner L, Ferdinand F-J, Friis RR, Bauer H, Pedersen NC, Johnson L, Theilen GH (1984) Biological and biochemical characterization of a new isolate of feline sarcoma virus: Theilen-Pedersen (TP1-FeSV). Virology 138: 324–331

Virus-Host Interactions and the Pathogenesis of Murine and Human Oncogenic Retroviruses*

P. N. Tsichlis[1] and P. A. Lazo[1,2]

1	Introduction	96
2	Oncogenesis by Nonacute Type C Retroviruses in Mice	97
3	Determinants of Viral Pathogenicity	98
3.1	The Virus	98
3.1.1	Generation of LTR and *env* Mutant and Recombinant Viruses During the Preleukemic Phase in Virus Infected Mice	98
3.1.2	The Role of the LTR in Oncogenesis	100
3.1.3	The Role of *env* and Other Viral Sequences in Oncogenesis	104
3.1.4	Genetic Variants with Altered Pathogenicity	106
3.1.5	Interactions Between Viruses	107
3.2	The Host	108
3.2.1	Host Genes Affecting Virus Replication	108
3.2.2	Genes Affecting the Host's Immune Response to Virus or Virus Infected Cells	111
3.2.3	Genes Affecting the Size of the Target Cell Pool	114
3.2.4	Genes Affecting Primarily the Type as Opposed to the Incidence of Disease	115
4	Insertional Mutagenesis	115
4.1	Effects of Provirus Insertion	117
4.1.1	Promoter Insertion	117
4.1.2	Enhancer Insertion	119
4.1.3	Truncation of the 5′ or the 3′ End of the Gene: Synthesis of an Abnormal Gene Product	119
4.1.4	Stability of the RNA Message	122
4.1.5	Long Distance Gene Activation	122
4.1.6	Activation of Multiple Neighboring Genes	123
4.1.7	Insertional Mutagenesis of Recessive Oncogenes: Gene Inactivation/Dominant Negative Mutations	123
4.1.8	Chromosomal Rearrangements by Homologous Recombination Between Integrated Proviruses	125
4.2	Genes Affected by Provirus Insertion	126
4.2.1	Growth Factors	126
4.2.2	Growth Factor Receptors	128
4.2.3	Membrane Associated or Cytoplasmic Proteins Involved in Signal Transduction	130
4.2.4	Nuclear Proteins/Transcription Factors	133
4.2.5	Partially Characterized Genes	137
4.2.6	Loci Marking Genes that Remain Undefined	138

* The work in PNT's laboratory is supported by Public Health Service grant CA-38047, and CA-06927, RR-05539 to the Fox Chase Cancer Center and an appropriation from the Commonwealth of Pennsylvania.
[1] Department of Medical Oncology, Fox Chase Cancer Center, Philadelphia, PA 19111, USA
[2] *Present address*: Unidad de Genetica Molecular, Centro Nacional de Biologia Celular y Retrovirus, Instituto de Salud "Carlos III", 28220 Majadahonda (Madrid), Spain

4.3   Spontaneous Recurrent Provirus Integration in Virus Infected Preleukemic
      and Leukemic Cell Clones 141
4.4   Cooperating Oncogenes 142

5   Transduction of Cellular Oncogenes 143

6   Transactivation of Cellular Genes by Viral Gene Products 144

7   Concluding Remarks 148

References 149

1 Introduction

Retroviruses are a heterogeneous group of plus strand RNA viruses which replicate their RNA genome through a DNA intermediate. During their life cycle the retroviruses interact with the host in a variety of ways. Thus, viral gene products interact with cellular proteins and modulate or subvert their function. Furthermore, the proviral DNA integrates into the cellular genome introducing mutations which alter the biology of the infected cell. These interactions between the virus and the host and the phenomena associated with them determine the pathogenicity of retroviruses (LAZO and TSICHLIS 1990 and this volume).

Based on their pathogenic potential, the retroviruses are divided into three subfamilies: *Oncovirinae* which are responsible mainly for the induction of neoplasms, *Lentivirinae* which have been associated with a variety of non-neoplastic diseases, and *Spumavirinae* which are not known to be pathogenic (TEICH 1982). The *Oncovirinae* are subdivided into two major groups: nonacute (crucial) transforming and acute transforming retroviruses (TEICH 1982). Viruses of this subfamily are primarily responsible for the development of neoplasms, although they also may be responsible for the development of non-neoplastic diseases. For example, chickens inoculated with avian leukosis virus (ALV) may develop osteopetrosis, a wasting syndrome or autoimmune phenomena; cats inoculated with feline leukemia virus (FeLV) may develop severe immuno-deficiency; and mice inoculated with a variety of retrovirus strains may develop anemia, lymphoproliferation, or neurologic defects. Finally, humans infected with HTLV-I may develop severe neurological manifestations such as tropical spastic paraparesis (TSP) (LAZO and TSICHLIS 1990).

In this review we will concentrate on the interactions between the virus and the host that lead to the development of neoplasms by nonacute rodent and human type C retroviruses of the subfamily of *Oncovirinae*. The mechanisms of oncogenesis by other members of the same subfamily, such as the avian and feline type C and murine type B retroviruses, are dealt with in detail in other chapters of this volume. Therefore, these viruses will be discussed only when necessary to illustrate the general concepts of retrovirus pathogenesis.

2 Oncogenesis by Nonacute Type C Retroviruses in Mice

Nonacute murine type C retroviruses induce neoplasms that are exclusively hematopoietic in origin. Tumor induction by these viruses is a complex process of continuous and dynamic evolution. For example, the process of tumor induction by Moloney murine leukemia virus (MoMuLV) and other murine type C retroviruses can be divided into four sequential stages (LAZO and TSICHLIS 1990). The first stage, detected within weeks following virus inoculation, is characterized by a polyclonal proliferation of hematopoietic precursor cells in the bone marrow and spleen of the inoculated animals. The second stage, which occurs within 40–50 days, is characterized by clonal expansion of cells in the spleen (TSICHLIS, unpublished data). The third stage is characterized by the appearance of T cell lymphomas, which arise in the thymus 3–6 months following virus inoculation. The last stage of tumor progression is characterized by the continuous selection of rapidly growing or metastatic cell clones.

The phenotype of T cell lymphomas induced by nonacute murine type C retroviruses is variable (LAZO et al. 1990a). This property is unique to these tumors and distinguishes them from other virus-induced tumors which have a fixed phenotype (SCHÜPBACH 1989; DEPELCHIN et al. 1989; RISSER 1982). The development of tumors with variable phenotypes may be due to one of two mechanisms: (a) Murine type C retroviruses infect and transform cells at all stages of T cell development. (b) These viruses transform immature T lymphocytes which undergo limited differentiation during oncogenesis. The notion of limited differentiation suggested by the second hypothesis is supported by several lines of evidence. We have shown that tumor cells are phenotypically unstable and have suggested that this is due to the differentiation of these cells (LAZO et al. 1990a). Furthermore, earlier studies suggest that the target cells for virus infection and transformation are immature T cells. According to these studies, virus inoculation in newborn animals is followed by the appearance of cells committed to neoplastic transformation in the bone marrow and spleen (ASJÖ et al. 1985). Since the majority of the cells in these organs are phenotypically immature (DAVIS et al. 1987; DAVIS et al. 1986; STORCH et al. 1985), these findings suggest that the partially transformed cells belong to an early stage in T cell development. In separate experiments, it has been shown that the great majority of the infected cells in young AKR mice inoculated with lymphomagentic mink cell focus-forming viruses are immature cortical thymocytes (O'DONNELL et al. 1984; CLOYD 1983). The immaturity of the target cells, combined with the phenotypic instability of the developing neoplasms, suggests that differentiation in the thymus may be an important component of the process of oncogenesis.

3 Determinants of Viral Pathogenicity

3.1 The Virus

Early studies to determine the mechanism(s) of retrovirus pathogenesis concentrated on the identification and characterization of the viral sequences responsible for pathogenicity. The first experiments along these lines were designed to determine the sequences responsible for the oncogenic potential of the ALVs. To this end, recombinants were generated between an oncogenic (td-Pr-RSV-B) and a nononcogenic (RAV-O) ALV. Inoculation of chickens with these recombinant retroviruses revealed that the main determinant of the oncogenic potential of the virus mapped within the viral long terminal repeat (LTR) (TSICHLIS and COFFIN 1979; ROBINSON et al. 1979; TSICHLIS and COFFIN 1980; TSICHLIS et al. 1982; LAIMINS et al. 1984b). Since the LTR contained only non-coding sequences, these experiments suggested that tumor induction by ALVs was due to insertional mutagenesis of the cellular genome mediated by provirus integration (TSICHLIS and COFFIN 1979; TSICHLIS and COFFIN 1980), a prediction which has been amply confirmed experimentally (this volume). Later experiments revealed that, in addition to the LTR, other viral sequences contributed to the oncogenicity of the virus (ROBINSON et al. 1982).

Subsequent detailed genetic studies of the murine type C retroviruses confirmed and extended these early observations. Using recombinant and mutant viruses constructed in vitro or isolated from preleukemic or leukemic animals, several groups of investigators have shown that transcriptional control sequences within the LTR influence both the oncogenic potential and the disease specificity of these viruses (CHATIS et al. 1983; CHATIS et al. 1984; DESGROSEILLERS and JOLICOEUR 1984; HOLLAND et al. 1989; ISHIMOTO et al. 1985, 1987; LENZ et al. 1984; LI et al. 1987). In addition, sequences outside the LTR, and more importantly in the *env* gene, contribute to the viral pathogenic potential (HARTLEY et al. 1977; CLOYD et al. 1980; THOMAS et al. 1986; MUCENSKI et al. 1987b; OLIFF et al. 1984, 1985; HOLLAND et al. 1985a). The following description of the role of viral genetic elements in oncogenesis will concentrate on the non-acute murine type C retroviruses.

3.1.1 Generation of LTR and *env* Mutant and Recombinant Viruses During the Preleukemic Phase in Virus Infected Mice

Early studies of viral gene expression in the lymphoid organs of high leukemia mouse strains showed a rapid increase of expression of viral antigens in the thymus during the late preleukemic period. This surge of viral gene expression was due to the appearance of a new class of recombinant viruses, which had an extended host range (HARTLEY et al. 1977) and were thymotropic (GREEN et al.

1980). These viruses were named either MCF (mink cell focus-forming viruses), because they formed cytopathic foci in mink cells (HARTLEY et al. 1977),or polytropic, because of their unique property to infect both mouse and hetero-logous cells in culture (HIAI et al. 1977). The molecular events responsible for the generaton of the MCF recombinant viruses, which have been the subject of intense study for more than a decade, were only recently clarified. In addition to the small number of endogenous ecotropic proviruses inherited by the majority of the laboratory mouse strains (JENKINS et al. 1982; STOYE and COFFIN 1985), recent studies permitted the identification of at least 40 xenotropic and 77 polytropic endogenous proviruses (BLATT et al. 1983; WEJMAN et al. 1984; O'NEIL et al. 1986; STOYE and COFFIN 1987, 1988; FRANKEL et al. 1989a; FRANKEL et al. 1989b; FRANKEL et al. 1990). Of the 77 members of the polytropic provirus family 30 have been classified in a distinct subfamily, the modified polytropic proviruses, which among other structural features exhibit a 27 bp deletion in the *env* gene (STOYE and COFFIN 1987, 1988; FRANKEL et al. 1990). Using strain distribution patterns in recombinant inbred strains, the position of the majority of these proviruses in the mouse genome has been determined (FRANKEL et al. 1990). The generation of MCF viruses involves two successive recombination events. The first of these events between an ecotropic virus and the xenotropic virus *Bxv*-1 on mouse chromosome 1 (FRANKEL et al. 1990) gives rise to an ecotropic virus with a new LTR derived from the xenotropic virus parent (CHATTOPADHYAY et al. 1981; KHAN 1984; QUINT et al. 1984; HOGAN et al. 1986). The second event substitutes the ecotropic with a polytropic *env* gene derived from multiple independent polytropic virus parents (FRANKEL et al. 1989b). One endogenous polytropic provirus, which has been mapped near *Fv*-1 on mouse chromosome 4 (*Pmv*-25), has been suggested as a preferred *env* gene donor on the basis of its strain distribution pattern among 13 AKXD recombinant inbred mouse strains; however, *Pmv*-25 is absent from the high leukemia strain C58/J, indicating that its presence is not necessary for the generation of MCF type recombinant retroviruses (FRANKEL et al. 1989b).

Several lines of evidence suggest that the MCF viruses play a critical role in T cell leukemogenesis. These include the timing of their appearance in the late preleukemic phase in high leukemia mouse strains and their reproducible isolation from retrovirus-induced murine T cell lymphomas (HARTLEY et al. 1977; HIAI et al. 1977). Recombinant inbred mice between a high leukemia (AKR) and a low leukemia (DBA/2) mouse strain develop T cell lymphomas with a high, low, or intermediate frequency. Inheritance of the loci responsible for the generation of MCF recombinant viruses, combined with the lack of inheritance of *Rmcf*[r], a gene which restricts MCF virus replication, are linked to the high leukemia phenotype (MUCENSKI et al. 1987b, 1988a). Finally, direct testing has shown that MCF viruses are thymotropic (GREEN et al. 1980) and oncogenic (CLOYD et al. 1980).

In addition to the MCF viruses generated by two successive recombination events replacing both the LTR and the *env* gene of the ecotropic virus parent (class I, MCF viruses), a second class of MCF recombinants (class II), are formed by a single recombination event, and carry an extensive replacement of the env gene only (THOMAS and COFFIN 1982; LUNG et al. 1983; ROWE et al. 1979). These viruses are not oncogenic, with the exception of the class II viruses isolated from CWD mice which have been associated with the induction of nonthymic lymphomas (THOMAS et al. 1986). Furthermore, the isolation of class II MCF viruses from preleukemic AKR mice has been interpreted to suggest that they have the potential to be intermediates in the stepwise generation of class I polytropic viruses (EVANS and MALIK 1987).

The generation of variant retroviruses with altered oncogenic potential during the preleukemic phase in high leukemia mouse strains is usually due to recombination among different classes of replicating endogenous viruses. Leukemogenic virus variants, however, may also be due to point mutations. For example, the SL3-3 virus isolated from AKR mice carries a point mutation in the enhancer core region which alters the transcriptional activity of the LTR in T cells (BORAL et al. 1989; LOSARDO et al. 1989).

3.1.2 The Role of the LTR in Oncogenesis

Extensive genetic studies on the role of the LTR in the pathogenesis of murine type C retroviruses confirmed and extended the original observations on the role of the LTR in the pathogenesis of the avian retroviruses (TSICHLIS and COFFIN 1979, 1988; ROBINSON et al. 1979; TSICHLIS et al. 1982; LAIMINS et al. 1984b). These studies revealed that transcriptional control sequences in the LTR determine not only the oncogenic potential of these viruses, but also the latency period of disease induction and the disease specificity (CHATIS et al. 1983; CHATIS et al. 1984; DESGROSEILLERS and JOLICOEUR 1984; HOLLAND et al. 1989; ISHIMOTO et al. 1985, 1987; LENZ et al. 1984; LI et al. 1987). To explain these findings it was hypothesized that the LTR was a determinant of the host range of the virus and that enhanced viral growth in the target organ, mediated by the transcriptional specificity of the LTR, may be a necessary and sufficient requirement for disease induction. This was strengthened by the early finding that thymomagenic recombinant viruses were also thymotropic (GREEN et al. 1980). To test this hypothesis the tropism and pathogenic potential of natural viral isolates and in vitro constructed recombinants were determined. These studies revealed that the virus tropism per se correlated only partially with leukemogenic potential and disease specificity, suggesting that although viral growth in the target organ may be necessary, it may not be sufficient for disease induction (DESGROSEILLERS et al. 1983a; ROSEN et al. 1985; EVANS and MORREY 1987; SHORT et al. 1987; HOLLAND et al. 1989). This may be because the LTR, which is the major determinant of the pathogenic potential of the virus (CHATIS et al. 1983, 1984; DESGROSEILLERS et al. 1983b; DESGROSEILLERS and JOLICOEUR 1984; ISHIMOTO et al. 1985, 1987; LENZ et al. 1984; LI et al. 1987; HOLLAND et al. 1989), is one of the determinants of the viral host range (DESGROSEILLERS et al. 1983a; ROSEN et al. 1905; HOLLAND et al. 1985a, b; EVANS and MORREY 1987; HOLLAND et al. 1989) but not the only one (EVANS and MORREY 1987; HOLLAND et al. 1989).

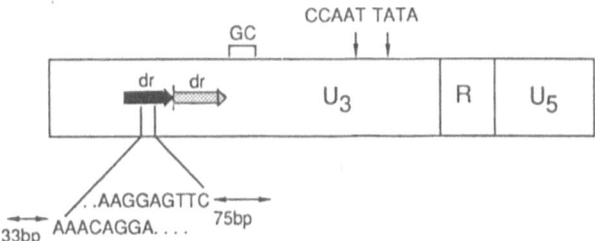

Fig. 1. Structure of the murine type C retrovirus LTR. The relative position of the GC rich domain (GC) and the CAT (CCAAT) and TATA boxes have been marked by *arrows* above the U_3 region. The *arrows* marked *dr* within the U_3 region identify the LTR direct repeats. The boundaries of the central portion of the dr which is repeated in the majority of virus isolates have been marked. These boundaries may extend up to 33 bp 5′ or 75 bp 3′ in individual virus isolates

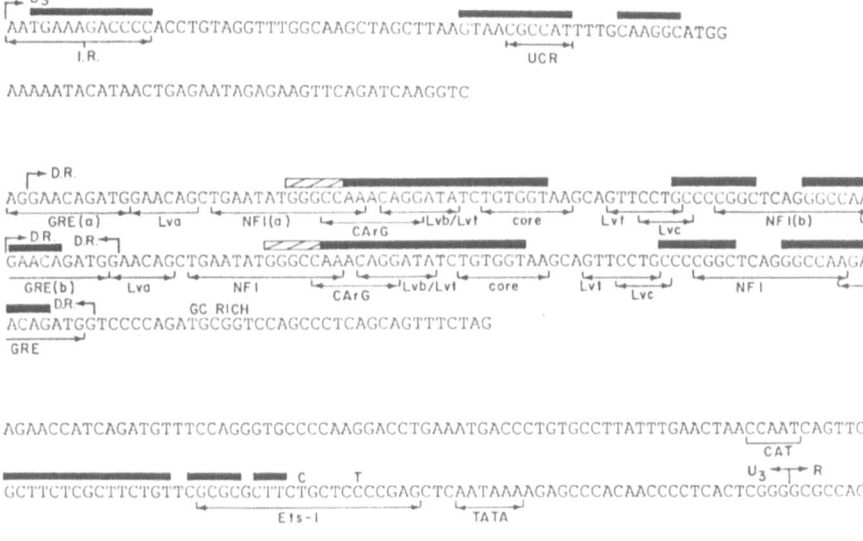

Fig. 2. Evolutionarily conserved and functional motifs of the MoMuLV LTR. The LTR sequence has been divided into three portions which are presented in three sequential panels. The *middle panel* contains the two direct repeats, whose boundaries are marked by *arrows*, and the GC rich region. The *bars* above the sequence identify evolutionarily conserved regions. The *hatched bars* identify sequences which are less conserved than the ones marked by *full bars. Arrows* under the sequence identify important regulatory regions such as the inverted repeat (IR), CAT and TATA boxes, polyadenylation signal, and transcription factor binding sites. Each direct repeat contains two GRE elements and two NF1 binding sites. To distinguish between them, the 5′ sites were marked *GRE(a)* and *NF1(a)* while the 3′ sites were marked *GRE(b)* and *NF1(b)* in the first direct repeat only. The MoMSV and the MoMuLV LTR differ by 2 bps in the *Ets*-1 binding region, as indicated

To explore the role of the LTR in oncogenesis, extensive attempts have been made over the last 5 years to dissect its function as a transcriptional regulator. Early studies revealed that the LTR of Moloney murine sarcoma virus (MoMSV), a prototype murine type C retrovirus, harbors at least two distinct positive activators of transcription: a distal signal located between the TATA and CAT boxes and an enhancer (GRAVES et al. 1985a, b). The two activators of transcription appeared to function independently. Thus, the enhancer was active in the absence of the distal signal in mouse L cells and the distal signal was active without the enhancer in frog oocytes (GRAVES et al. 1985a). In addition, infection of mouse L cells with herpes simplex virus (HSV) enhanced the activity of the distal signal but did not affect the enhancer (GRAVES et al. 1985b). Structural and functional analysis of the LTR to date has identified a series of critical elements (Figs. 1 and 2). These include the inverted repeats (IR) involved in integration the upstream control region (UCR), a negative transcriptional regulator, the direct repeats (dr) containing the viral enhancer, the GC rich sequence, the CAT box, the Ets-1 binding site and the TATA box. The sequences that are repeated in the context of the dr vary among viruses (GOLEMIS et al. 1990) (Fig.1).

Some sections of the viral LTR are highly conserved among viruses while other sections vary considerably (GOLEMIS et al. 1990) (Fig. 2). The conserved sequences appear to identify regions of functional significance that may be critical for the viability of the retrovirus genome. The variable regions may allow for the unique features of each virus and the rich biological characteristic of retrovirus infection.

The Viral Enhancer

The LTR of naturally occurring type C retrovirus isolates usually contains two or more copies of a tandem repeat with enhancer activity whose boundaries vary among viruses (Fig. 1) (GOLEMIS et al. 1990). The enhancer repeat in the majority of virus isolates contains a highly conserved motif AAACAGGATATCTG(T/C)GGT which is flanked to its 5' side by the less conserved sequence GGGCC (Fig. 2). The central portion of this sequence (CAGGTA) has been recognized as a binding site for the leukemia virus factors b and t (Lvb/Lvt) (SPECK and BALTIMORE 1987). The 3'' portion (TGTGGTAA in MoMuLV) corresponds to the enhancer core motif (TGTGG(T/A)(T/A)(T/A)) originally identified in the SV40 and the polyoma virus enhancers (WEIHER et al. 1983). Several lines of evidence suggest that these motifs are critical in determining viral tissue tropism, oncogenic potential, and disease specificity of individual virus isolates. Mutations in the Lvb/Lvt binding site decreased transcription from the MoMuLV enhancer in all cell types while mutations in the core region affected the activity of the enhancer only in lymphoid cells (SPECK et al. 1990b). Similarly, a single base pair difference between the SL3-3 and Akv virus in the core region appears to be responsible for the enhanced activity of the SL3-3 enhancer in T cells (BORAL et al. 1989; LOSARDO et al. 1989), although this effect could be detected only in the context of additional upstream sequences

(LoSARDO et al., in press). Sequence comparisons originally revealed differences between the core motifs of erythroleukemia and T cell leukemia inducing viruses suggesting that these elements may be involved in determining disease specificity (CLARK et al. 1985). This was confirmed by more recent studies which showed that MoMuLV mutants in both copies of the core motif(and to a lesser degree in both copies of the Lvb/Lvt motif) induce erythroleukemias after an extended latency period as opposed to the wild-type virus which induces solely T cell lymphomas following a relatively short incubation (SPECK et al. 1990b).

There are two additional conserved regions 3' of the Lvb/Lvt and core motifs which contain binding sites for nuclear factor-1 (NF-1) and the glucocorticoid receptor (GRE) (Fig. 2). Both sites contribute to the transcriptional activity of the enhancer. Thus, mutations in the GRE elements affect transcription in all cell types while mutations in the NF-1 site affect transcription primarily in fibroblasts (SPECK et al. 1990a). The region, including the Lvb/Lvt, core, and NF-1 binding sites and the flanking GRE element, is a component of the enhancer in most viruses and it has been designated the 'enhancer framework' region. A 30 bp region overlapping with the 5' end of the enhancer framework in the MoMuLV LTR appears to regulate LTR esponsiveness to phorbol ester (TPA) in Jurkat cells (SPECK et al. 1990a). It has been proposed that this phenomenon is regulated by the motif CCAAACAGG (Fig. 2), which is similar to the CArG box (GOLEMIS et al. 1990) found in the central portion of the serum response element and binding the p67 serum reponse factor (TREISMAN 1985, 1986).

Immediately 3' of the "enhancer framework" there is a 35 bp long GC rich region which is included in the LTR dr in some viruses and which also exhibits enhancer activity (LAIMINS et al. 1984a). The nondefective Friend murine leukemia virus (F-MuLV) clone 57, which induces erythroleukemias, and MoMuLV, which induces T cell lymphomas, exhibit significant differences in the GC rich domain, while they are identical in the Lvb/Lvt/core region (GOLEMIS et al. 1990). To determine whether the GC rich region contained element responsible for disease specificity, recombinants within this segment were constructed between F-MuLV and MoMuLV and tested for their disease inducing phenotypes. The results revealed that the 5' half of this region in F-MuLV contains a determinant that contributes to the ability of this virus to induce erythroleukemia (GOLEMIS et al. 1989).

In addition to the sites we have already discussed, several more factor binding sites can be recognized within the enhancer region of MoMuLV and other related viruses. These include the GRE (a), NF-1(a), Lva, and Lvc sites (SPECK and BALTIMORE 1987; GOLEMIS et al. 1990) (Fig. 2). Mutations in both copies of the Lva and Lvc sites or the promoter proximal GRE element did not affect the oncogenic potential of MoMuLV. On the contrary, mutations affecting two or three GREs, both copies of the Lvb/Lvt site, two of the four NF- 1 sites, or both copies of the enhancer core region delay significantly the induction of disease by MoMuLV. In addition, as stated earlier, mutations in the core or Lvb/Lvt motifs also affect the disease specificity of the virus (SPECK et al. 1990b). Recent studies have shown that, in addition to the common factors binding the

MoMuLV and the F-MuLV enhancer regions, several more factors bind uniquely to each enhancer (SPECK and BALTIMORE 1987; MANLEY et al. 1989). Although not fully explored to date, these differences may be critical for determining the disease specificity of each virus.

Recently, several laboratories have concentrated on the factors that bind the enhancer core, a region that plays a critical role in determining disease specificity (THORNEL et al. 1988; BORAL et al. 1989; JOHNSON et al. 1987). Studies utilizing the Akv and SL3-3 core regions identified three binding factors: S-CBF which binds only the SL3-3 core, A-CBF which binds only the Akv core, and S/A-CBF which binds the core motifs of both viruses. Of these factors S-CBF appears to be critical in the ability of the cell to distinguish the Akv from the SL3-3 core. When this factor is missing both core motifs exhibit comparable enhancer activities (BORAL et al. 1989). Preliminary data have suggested that expression of these factors in T cell lines may depend on the stage of cell differentiation (J.LENZ; personal communication).

Promoter Associated Motifs

In addition to the CCAAT and TATA boxes, two more conserved domains have been identified in the promoter region. Both are located between the CCAAT and the TATA motifs (Fig. 2). Studies utilizing the LTR of MoMSV have identified a 20 bp region immediately upstream from the TATA box which binds specifically the product of the Ets-1 proto-oncogene and whose 5' half is contained within one of these two conserved domains (Fig. 2). A MoMSV LTR construct containing a disrupted Ets-1 binding site exhibits decreased transcriptional activity in mouse T lymphocytes in culture, in comparison to the same construct containing the wild-type binding site (GUNTHER et al. 1990).

Upstream Control Region

In addition to the inverted repeat, two more conserved domains were identified in the LTR region upstream from the enhancer (Fig. 2). A motif, CGCCAT, included within the 5' conserved domain was recently identified as a negative control element and was named UCR (FLANAGAN et al. 1989). At least two nuclear factors, UCR-U and UCR-L, bind within this element. The role of this element as a negative regulator of transcription is supported by several lines of evidence: (a) MuLV transcription in mouse splenocytes correlates inversely with the detection of nuclear proteins binding the UCR; (b) Treatment of mice with lipopolysaccharide is associated with enhanced MuLV expression and disappearance of UCR binding proteins in the spleen; and (c) UCR oligonucleotides enhanced the expression of MuLV LTR/CAT constructs transfected in mouse L cells, presumably by competing for L cell inhibitory factors (FLANAGAN et al. 1989).

3.1.3 The Role of *env* and Other Viral Sequences in Oncogenesis

The role of the *env* gene in oncogenesis, which was originally suggested by the isolation of MCF viruses (HARTLEY et al. 1977; HIAI et al. 1977), was confirmed by the direct testing of the oncogenic potential of naturally isolated (CLOYD et al.

1980) or in vitro constructed envelope recombinants (OLIFF et al. 1984; HOLLAND et al. 1985a). Several mechanisms have been proposed to date to explain how the product of the *env* gene may be involved in the process of oncogenesis: First, a virus with a novel *env* gene may circumbent superinfection resistance by ecotropic viruses achieving efficient infection of virus infected cells. Such cells, representing the putative progenitors of the tumor cells, proliferate actively in the spleen in the early stages of virus infection. Their efficient infection by MCF viruses arising in the thymus may be the critical step responsible for cellular transformation via insertional mutagenesis (DAVIS et al. 1987). Second, simultaneous infection by ecotropic and MCF viruses may suppress the growth of bone marrow stromal cells, which in turn inhibits hematopoiesis. According to this hypothesis, the proliferation of hematopoietic cells in the spleen in the early stages of oncogenesis is a manifestation of compensatory extramedullary hematopoiesis. The inhibitory effects of viral infection on the growth of bone marrow stromal cells may be secondary to the masking of critical growth factor receptors by the attachment of the viral envelope gene product (BRIGHTMAN et al. 1990). Third, interaction of the MCF envelope gene product with its receptor may stimulate cell growth. Early experiments had suggested that individual tumor cells are infected by and produce MCF viruses whose *env* gene product binds to the tumor cell surface and may induce cellular proliferation (WEISSMAN and MCGRATH 1982). More recent studies have suggested the existence of rare T cell receptor-L3T4 complexes on the surface of retrovirus-induced lymphoma cells and have provided evidence for binding of the cognate retrovirus to these complexes. Cell proliferation is inhibited by antibody mediated blocking of this interaction (O'NEILL et al. 1987). Based on observations that gp55, the defective envelope glycoprotein of the spleen focus-forming virus (SFFV), interacts with the erythropoietin receptor (LI et al. 1990), studies have been initiated to determine whether the receptor utilized by the MCF envelope is related to the receptor of any of the known hematopoietic growth factors. Our ongoing studies suggest that the MCF envelope may affect the proliferative capacity of virus infected tumor cells, perhaps by interacting with the interleukin-2 receptor (TSICHLIS and BEAR 1991). To date it is not clear which one of these proposed mechanisms plays the major role in oncogenesis by MCF recombinant viruses. Since all these mechanisms, however, are supported by experimental data, it is likely that the role of the MCF envelope in oncogenesis is mediated by a complex and variable interplay between them.

During virus infection the product of the *env* gene is processed by a cellular protease into two proteins which, in the case of the murine type C viruses, are the extracellular protein gp70, which binds the viral receptor, and the transmembrane protein p15E. The sequence at the cleavage site separating gp70 from p15E is highly conserved among viruses (DICKSON et al. 1985). The two proteins are linked together by disulfide bonds. Most of the studies we discussed address the potential role of gp70 in oncogenesis. The transmembrane protein p15E, however, may also be involved in this process since several studies have suggested that it may exert a generalized immunosuppressive effect on virus infected hosts (RUEGG et al. 1989).

In addition to the viral LTR and the viral *env* gene, other viral genetic elements may also be involved in the process of oncogenesis. This has been suggested by studies on the oncogenic potential of numerous "in vitro" constructed recombinant retroviruses (OLIFF et al. 1985). Since we have no clues to date of the potential mechanisms involved in oncogenesis by these viruses, we will not discuss them further.

3.1.4 Genetic Variants with Altered Pathogenicity

Three independent enzyme systems are involved in the replication of the retroviral genome: (1) reverse transcriptase which generates a DNA copy of the viral RNA genome; (2) cellular DNA polymerase which replicates integrated proviral DNA sequences; and (3) RNA polymerase II which transcribes the integrated provirus. Two of these enzymes, reverse transcriptase and RNA polymerase II, lack proofreading properties. This leads to a high error rate which is responsible for the accumulation of mutations. The rate of mutation has been estimated from different experiments to be $1 \times 10^{-4} - 2 \times 10^{-5}$ per generation (DOUGHERTY and TEMIN 1988; COFFIN et al. 1980; COFFIN 1986). The mutations observed in replicating retrovirus genomes include point mutations and rearrangements such as deletions, duplications, or inversions of viral sequences. Additional processes responsible for the generation of genetic heterogeneity include recombination events between replicating retroviral genomes (COFFIN 1979) or between retroviral and cellular sequences (BISHOP and VARMUS 1985).

It has been estimated that at least half of the replicating viral genomes generated during a single replication cycle are mutated (DOUGHERTY and TEMIN 1988). This indicates that after repeated cycles of replication no two retroviral genomes will be identical. This has indeed been shown in the case of HIV. The *tat* gene of this virus was shown to be completely heterogeneous in sequence in multiple viral isolates from a single infected individual (MEYERHAUS et al. 1989). The inherent viral instability combined with the selective pressures imposed on the replicating retroviral genomes allows for the selection of retrovirus variants with unique properties such as enhanced or altered pathogenicity.

Examples of genetic variants with enhanced pathogenic potential include the acute transforming viruses, which are generated by recombination between viral sequences and cellular proto-oncogenes (BISHOP and VARMUS 1985 and this review, Sect. 5), and the mutant and recombinant viruses generated during the preleukemic phase in high leukemia mouse strains (TEICH et al. 1982 and this review, Sect. 3.1.1). Viruses with altered pathogenicity include immunosuppressive and neurotropic variants isolated in the course of retroviral infections. Characterized viral mutants isolated in the course of viral infection in mice include the SFFV component of the Friend leukemia virus complex (TEICH et al. 1982), the virus Du5H/LP-BM5, which induces a mouse immunodeficiency syndrome (MAIDS) (AZIZ et al. 1989; MORSE et al. 1989), the neurotropic temperature-sensitive MoMuLV mutants *ts*-1, *ts*-7, *ts*-11 (MCCARTER et al. 1977;

WONG et al. 1983), MoBA-I (BILLELO et al. 1986), and a neurotropic mutant isolated following passage of F-MuLV in rats (KAI and FURATA 1984). In addition, naturally selected virus isolates may exhibit characteristic properties that distinguish them from other related retroviruses. Thus, the ecotropic virus Cas-Br-E, isolated from wild mice trapped in the Lake Casitas area, in addition to inducing a variety of hematopoietic neoplasms, also induces a spongiform encephalopathy with high frequency (GARDNER et al. 1973; GARDNER 1985).

The mechanism by which these virus isolates induce disease is diverse and not always well understood. Thus, SFFV induces rapid erythroleukemia by direct interaction between the defective viral envelope glycoprotein gp55 and the erythropoietin receptor (LI et al. 1990). The MAIDS virus induces severe lymphoid hyperplasia involving certain T lymphocyte subsets (MORSE et al. 1989). The viral genetic determinant responsible for the lymphoproliferative response has been mapped within the *gag* gene (AZIZ et al. 1989). Finally, the neurotropism of both the ts mutants of MoMuLV (WONG 1990) and the Cas-Br-E virus isolate (DESGROSEILLERS et al. 1984) has been mapped within the *env* gene. Again, the mechanism of neurotoxicity by the *env* gene product remains unknown. In the case of the MoMuLV ts mutants it has been thought that, since the virus is defective in the proteolytic processing of the envelope gene product, accumulation of the precursor polyprotein may be toxic to infected glial and neuronal endothelial cells (WONG 1990). However, this is not the case in the Cas-Br-E virus isolate which is not defective in the processing of the viral envelope.

The induction of neurological damage by the neurotropic viruses depends on efficient virus replication. Thus, all retrovirus constructs containing a common envelope gene but different LTRs are neurotropic, but they induce different neurologic syndromes because the cells they infect may be differently distributed within the nervous system (DESGROSEILLERS et al. 1985).

3.1.5 Interactions Between Viruses

Two viruses, related or unrelated and replicating in the same cell, may interact in a variety of ways. These interactions may modify the biological properties and pathogenic potential of both. The modification may be phenotypic, due to pseudotyping or transactivation,or genotypic, due to recombination between the viral genomes. To describe these interactions we will not concentrate exclusively on the murine type C retroviruses. This is because some of these interactions can be illustrated better in avian or human retroviruses. Furthermore, since all retroviruses share common properties, the rules of interaction may be generally applicable to all of them.

Pseudotyping or phenotypic mixing, i.e., the exchange of envelope glycoproteins between related or unrelated viruses, has been known for many years. Thus, it has been shown that avian, murine, and human retroviruses may exchange envelope glycoproteins with vesicular stomatitis virus (VSV), a rhabdovirus (Weiss 1980). Similarly, subgroups of avian or murine retroviruses replicating in the same cell may exchange envelope glycoproteins, thus altering their host range (Vogt 1967; Ishimoto et al. 1977). Along the same lines, it was observed recently that human retroviruses may undergo phenotypic mixing with murine retroviruses (Lusso et al. 1990; Bacon et al. 1989, Lusso et al. 1989; Spector et al. 1990). Phenotypic mixing may alter the host range of retroviruses across species, between members of a single species, or among cell types in a single infected individual. Such alterations in host range may have profound effects on pathogenesis.

Transactivation between viruses has been studied extensively because of its potential role in the pathogenesis of human retroviruses. Thus, it has been shown that several non-HIV coded proteins, such as the adenovirus EIA 13S protein (Rice and Mathews 1988), the HSV-1 ICPO protein (Mosca et al. 1987), the CMV 1E2 protein (Davies et al. 1987), the X protein of HBV (Seto et al. 1988), the tax protein of HTLV (Yoshida and Seiki 1987), and an unknown protein product of herpes virus type 6 (Horvat et al. 1989), can transactivate HIV. Along the same lines, the HIV *rev* gene codes for a nuclear phosphoprotein that can replace functionally the HTLV *rex* gene product (Rimsky et al. 1988).

Recombination between the genomes of unrelated viruses is a well known phenomenon. The classical example is the recombination between adenovirus and SV40, which was described about three decades ago (reviewed by Grodzicker 1980). Most recently it was shown that retroviruses, which possess a very specialized recombination machinery, also have the ability to recombine with other viruses. Thus, it has been shown that avian retroviruses may integrate into the genome of the Marker's disease virus, a herpsvirus that causes a T cell lymphoproliferative disease in chickens (H.-J. Kung, personal communication). This kind of interaction may extend the host range and the pathogenic potential of retroviruses.

3.2 The Host

The development of hematopoietic neoplasms in retrovirus infected mice is under the control of multiple genetic loci (Table 1). Some of these loci control virus growth by affecting either virus replication or the host's immune response against the virus or virus infected cells. Other loci influence the process of oncogenesis by affecting the number of available target cells. Finally, additional loci affect the type of disease induced by the virus, by mechanisms that are not well understood. In the following paragraphs we will describe briefly our current understanding on the nature and function of these loci.

3.2.1 Host Genes Affecting Virus Replication

Susceptibility to virus growth in vivo is not sufficient for tumor induction. Thus, susceptible strains of mice inoculated with an ecotropic MuLV at 6 weeks of age

Table 1. Host genes affecting susceptibility to MuLV-induced disease

Genes affecting virus replication	
Provirus integration	*Fv-1*
Viral attachment and entry into the target cells	*Fv-4/Akrv*-1
	Fv-6/Rmcf
	Gv-1 and *Gv*-2
Other	*Srv*-1 and *Rgv*-2
Genes affecting the host immune response to the virus or virus infected cells	
MHC linked genes	*Rgv*-1
	Rfv-1 and *Rfv*-2
	Other
Genes not linked to MHC	*hr*
	Rfv-3
	X-linked immunodeficiency
	Fv-3
Genes affecting the size of the target cell pool	*Fv*-2
	W and *Sl*
	f
	nu[a]
Genes affecting primarily the type as opposed to the incidence of disease	*Fv*-5

[a] This gene may also affect the host immune response

develop significant viremia but they exhibit low tumor incidence. The same strains of mice inoculated at birth establish a similar degree of viremia and they proceed to develop tumors (BEAR and TSICHLIS , unpublished). Similarly, DBA/2 and A/J mice inoculated at birth develop significant viremia but they do not develop tumors (DEBRÉ et al. 1979). It should be pointed out, however, that although virus growth is not sufficient, it is necessary for tumor induction.

Host Genes Affecting Provirus Integration: The Fv-1 Locus
The *Fv*-1 locus on mouse chromosome 4 was originally identified as a locus that controls resistance to leukemogenesis by the Friend murine leukemia virus complex. Later studies showed that the *Fv*-1 locus controlled leukemogenesis by affecting the resistance to infection by most replication competent MuLVs. Furthermore, the *Fv*-1 mediated resistance to infection could be detected both in vivo and in culture (MERUELO and BACH 1983; PINCUS et al. 1971 a, b; ROWE et al. 1973; KOZAK 1985). Four alleles of the *Fv*-1 locus selectively restrict the growth of three viral genotypes (Table 2). Restriction to viral growth is dominant (MERUELO and BACH 1983). The resistance to retrovirus infection mediated by this locus is characterized by the accumulation of unintegrated linear proviral DNA in the nucleus of infected cells suggesting a defect in the process of proviral DNA integration (JOLICOEUR and RASSART 1981). However, it is not known whether the integration defect is due to the lack of a cellular cofactor required for integration or to the synthesis of defective proviral DNA. Mapping of the viral determinant of

Table 2. Alleles of the *Fv*-1 locus

Allele	Distribution	Restriction[a]	References
$Fv\text{-}1^n$	Inbred mice	Restricts the growth of B-tropic viruses	MERUELO and BACH 1983
$Fv\text{-}1^b$	Inbred mice	Restricts the growth of N-tropic viruses	MERUELO and BACH 1983
$Fv\text{-}1^{nr}$	Few inbred strains and wild mice	Restricts the growth of B-tropic and some N-tropic viruses	KOZAK 1985
$Fv\text{-}1^o$	Wild mice	No restriction	KOZAK 1985

[a] NB tropic viruses, selected by passage of B tropic viruses through $Fv\text{-}1^{n/n}$ cells, are not restricted by any of the $Fv\text{-}1$ alleles

Fv-1 tropism within the gene coding for the viral capsid protein (DESGROSEILLERS and JOLICOEUR 1983) has not provided additional clues about the molecular function of this locus.

Host Genes Affecting Viral Attachment and Entry into the Target Cells
Expression of viral envelope glycoproteins on the cell surface may interfere with virus infection by affecting viral attachment and entry into the target cells. At least four loci have been recognized to date which control expression of ecotropic or MCF type envelope glycoproteins on the surface of a variety of cell types in vivo (Table 1).

Fv-4/Akvr-1: The *Fv*-4 locus on mouse chromosome 12 is characterized by two alleles: *Fv*-4^r which is responsible for resistance to infection by ecotropic MuLVs and *Fv*-4^s which is responsible for the susceptible phenotype. The *Fv*- 4^r allele was originally recognized in FRG mice (SUZUKI 1975; SUZUKI et al. 1981). Later, it was also found in the Japanese wild mouse *Mus musculus molossinus* (ODAKA et al. 1978, 1901) and in California wild mice (GARDNER et al. 1980). The *Fv*-4^r allele is characterized by the insertion of a defective ecotropic provirus in mouse chromosome 12 (IKEDA et al. 1985; IKEDA and SUGIMURA 1989). This provirus lacks 5′ LTR and *gag* sequences and it contains a truncated *pol* gene, an ecotropic envelope gene, and a 3′ LTR. Expression of the *env* gene from a neighboring cellular promoter is responsible for viral interference and resistance to infection by ecotropic MuLVs (IKEDA and SUGIMURA 1989). Resistance is the dominant phenotype (GARDNER et al. 1980).

Fv-6/Rmcf: The resistance allele of this locus (*Rmcf*^{r}), which maps to mouse chromosome 5, is characterized by the expression of MCF envelope glycoproteins on the surface of hematopoietic cells (RUSCETTI et al. 1981; BASSIN et al. 1982; BULLER et al. 1987; FRANKEL et al. 1990). This is responsible for viral interference and resistance to infection by MCF viruses (RUSCETTI et al. 1981). The *Rmcf*[r] allele was originally recognized in DBA/2 mice which were found to be resistant to F-MuLV-induced leukemogenesis (RUSCETTI et al. 1981). It was originally thought that the *Rmcf*[r] allele was due to the inheritance of an MCF provirus (*Pmv*-40) in resistant mice. However, it was recently shown that, among 25 AKXD recombinant inbred mouse strains, two strains underwent recombination between the *Pmv*-40 provirus and the *Rmcf*[r] allele. Furthermore, NFSXDBA/2 *Rmcf*[r] congenic mice have been constructed which lack the *Pmv*-40 provirus (FRANKEL et al. 1989). These data suggest that the *Fv*- 6/*Rmcf* locus contains a regulatory gene which controls expression of MCF envelope glycoproteins in hematopoietic cells. The resistance phenotype is dominant.

Gv-1 and Gv-2 : These loci were originally defined by their effect on the expression of the GIX envelope antigenic determinant on the surface of differentiating thymocytes in the 129 mouse strain (STOCKERT et al. 1971). Existing evidence today suggests that at least one of them (*Gv-1*) is a regulatory locus that controls expression of endogenous retroviruses (STOCKERT et al. 1975; LEVY et al. 1985).

At least two additional loci that control resistance to the replication of murine retroviruses, *Srv*-1 and *Rgv*-2, have been described (TEICH et al. 1982). The mechanism of action of these loci, however, remains undetermined.

3.2.2 Genes Affecting the Host's Immune Response to Virus or Virus Infected Cells

Virus growth in vivo is affected not only by genes that influence viral replication, but also by genes that affect the host's immune response to the virus or virus-infected cells. Some but not all of these genes are linked to the major histocompatibility complex (MHC). The mechanism of their action is variable.

MHC Linked Genes
In 1964 Lilly et al. showed that the H-2^k haplotype of the MHC complex conferred susceptibility while the H-2^b haplotype conferred resistance to leukemogenesis by Gross passage A virus. The H-2 linked gene associated with resistance was named *Rgv*-1 and it was mapped in the I region close to the K end of the H-2 complex (LILLY et al. 1964; LILLY and PINCUS 1973). Later studies revealed that H-2 linked genes control the host's immune response against the virus or virus infected cells.

The host's immune response to virus infection may be characterized by the production of antibodies to the virus or virus infected cells. This type of response is mediated by class II MHC genes (BEAR et al. 1980; VASMEL et al. 1988). Other types of immune responses include the induction of lymphoid cell proliferation and the appearance of cytotoxic T cells which are directed against virus infected cells. These types of responses are mediated by class I MHC genes and they have been observed in AKR mice and in mice infected by the Friend virus complex and MoMuLV (GREEN 1983; PLATA and LILLY 1979; PLATA 1982; CHESEBRO et al. 1974; CHESEBRO and WEHRLY 1978; BRITT and CHESEBRO 1983; CHESEBRO et al. 1990; DEBRÉ et al. 1979).

MHC mediated immune responses may be enhanced by increasing the expression of MHC genes in the immunocompetent cells. This has been observed in mice of the H-2^b or H-2^d haplotype inoculated with radiation leukemia virus (RadLV) (MERUELO 1979; MERUELO et al. 1977). Since interferon-γ enhances the expression of genes of the H-2 complex (KING and JONES 1983; WONG et al. 1984), it is possible that this phenomenon may be due to interferon-γ production.

To determine the potential role of genes that map within the I region in leukemogenesis (Rgv-1), mice from susceptible and resistant strains were inoculated either at birth or at 6 weeks of age with an ecotropic MuLV (TUCKER et al. 1977; TSICHLIS et al. 1979; BEAR et al. 1980). Individual mice were scored for virus growth and tumor induction. All mice were susceptible to virus infection when inoculated at birth. High virus titers were detected in hematopoietic organs and they persisted through the animal's lifetime (TSICHLIS et al. 1979; BEAR et al. 1980). Lymphoid cell tumors originating in the spleen or mesenteric lymph nodes were detected with equal frequency in mice of all H-2 haplotypes within 12–18 months following virus inoculation (S.E. BEAR and P.N. TSICHLIS, unpublished). When inoculated at 6 weeks of age, certain mouse strains were resistant to retrovirus infection and the resistance correlated with the H-2 haplotype (TUCKER et al. 1977; TSICHLIS et al. 1979). The resistance gene(s) mapped within the I region of the H-2 complex. The resistance phenotype developed within the first 10 days of life, was T cell dependent, and correlated with the appearance of neutralizing antibodies in the serum of virus infected animals (BEAR et al. 1980). Since animals inoculated at birth continue to exhibit high virus titers in hematopoietic organs throughout life, it appears that early exposure to the virus abrogates the ability of the immune system to control ecotropic virus infection later in life. More recent experiments have suggested that the abrogation of the immune response against the virus in mice inoculated at birth may be unique to mice inoculated with ecotropic viruses. Mice carrying the resistant H-2^b haplotype are susceptible to perinatal infection by the polytropic virus MCF1233. However, 6 weeks later, resistance to MuLV infection becomes established and the virus titer drops. More significantly the drop of the virus titer correlates with resistance to the development of T cell lymphomas (VASMEL et al. 1988). The gene(s) regulating resistance to MCF1233 map also within the I region of the H-2 complex, and the resistance correlates with the appearance of antiviral antibodies in the serum of virus inoculated animals (VASMEL et al. 1988). The most likely explanation for these data is that a single immune response gene (Rgv-1) may control the generation of different immune responses to different viruses. An alternate but less likely possibility is that the immune response to different viruses and the ultimate outcome of the retrovirus infection may depend on an array of tightly linked genes unique for each virus.

Table 3. H-2 linked loci associated with resistance to viral leukemogenesis

Locus	Virus	Resistant haplotype	Map position within the H-2 complex	Mode of inheritance	References
Rgv-1	Gross passage A	H-2^b	K, I-A, I-J, I-E	Resistance dominant	LILLY and PINCUS 1973
Rfv-1	Friend MuLV	H-2^b	D	Susceptibility dominant	CHESEBRO et al. 1990
Other[a]					
Rfv-2	Friend MuLV	HK-2^b	K, I-A, I-J, I-E	Resistance dominant	CHESEBRO et al. 1990
Rmv-1	MoMuLV	H-2^b, H-2^s	K, I-A	?	DEBRÉ et al. 1980
Rmv-2	MoMuLV	H-2^b, H-2^s	I-E, S	?	DEBRÉ et al. 1980
Rmv-3	MoMuLV	H-2^b, H-2^s	D	?	DEBRÉ et al. 1980
Rrv-1	A-RadLV	H-2^s	I-A	?	LONAI et al. 1981
Rrv-2	A-RadLV	H-2^s	I-E, S	?	LONAI et al. 1981

[a] These may not be independent loci but variants of Rgv-1 and Rfv-1. Additional unnamed loci which can be classified here include: loci affecting susceptibility to the D, Kaplan, and RS strains of RadLV and loci affecting susceptibility to the MCF 1233 and Tennant leukemia viruses (for review see ZIJLSTRA and MELIEF 1986)

Other H-2 linked genes that determine resistance to viral leukemogenesis are listed in Table 3. One of them, Rfv -1, requires special mention. This gene controls recovery from leukemic splenomegaly in mice inoculated with the Friend virus complex and it has been mapped at the D end of the H-2 complex (CHESEBRO et al. 1974; CHESEBRO and WEHRLY 1978; BRITT and CHESEBRO 1983; CHESEBRO et al. 1990). The mapping of this gene suggests that it may be involved in the generation of cytotoxic T cell (CTL) responses against virus infected cells. Thus, it has been postulated that H-2D in mice carrying the H-2^b resistant haplotype may undergo preferential association with viral antigens to provide a stronger CTL recognition (CHESEBRO et al. 1990). To observe the full effect of Rfv-1 in recovery from Friend virus-induced splenomegaly, the Rfv-1^r allele requires the presence of resistant Rfv-2 (Tables 1 and 3) and Rfv-3 alleles (Table 1) (CHESEBRO et al. 1990).

Genes Not Linked to the MHC
Hr: This is an autosomal recessive mutation caused by the integration of an endogeneous provirus in mouse chromosome 14 (STOYE et al. 1988b). Phenotypically, it is characterized by decreased immune responses to T cell dependent antigens (MORRISSEY et al. 1980), detection of MCF viruses in the hematopoietic organs of mutant mice, and development of T cell lymphomas (MEIER et al. 1969; HIAI et al. 1977; GREEN et al. 1980). More recent studies have not confirmed the earlier observations that the incidence of viremia and T cell leukemia is significantly higher in hr/hr as opposed to hr/+ mice (MUCENSKI et al. 1988a). This suggests that the hr mutation and the susceptibility to leukemia induction

may be linked traits and that the recently observed lack of correlation between these traits may be due to genetic recombination (MUCENSKI et al. 1988a).

Rfv-3: This is a dominant non-H-2 linked gene that influences recovery from viremia in Friend virus inoculated mice. C57Bl/10 mice carry the resistant allele of this locus (*Rfv-3^r*) while A·BY mice carry the susceptible allele (*Rfv-3^s*) (CHESEBRO and WEHRLY 1979; CHESEBRO et al. 1990). The recovery from viremia mediated by the *Rfv*-3 locus correlates with the appearance of neutralizing antiviral antibodies in the serum of virus inoculated mice, suggesting that *Rfv*-3 influences antiviral immunity (DOIG and CHESEBRO 1979). The *Rfv*-3 locus appears to cooperate with other loci within the H-2 complex. Thus, *Rfv*- 3^r mice carrying the H-2^b haplotype not only recover from viremia, but they also recover from splenomegaly and leukemia induced by the Friend virus complex. This is not observed in mice with other H-2 haplotypes (CHESEBRO and WEHRLY 1979; CHESEBRO et al. 1990).

X-linked Immunodeficiency: The early stages of MoMuLV infection in mice are characterized by a polyclonal proliferation of Lyt 1$^+$2$^-$, 20α hydroxy-steroid dehydrogenase (20αSDH) positive cells (LEE and IHLE 1981a; IHLE et al. 1982; FUNG et al. 1984). This phenomenon is observed in CBA/J but not in the closely related CBA/N mouse strain which carries a recessive X-linked defect preventing the immune response to certain antigens. As a result CBA/N mice are relatively resistant to leukemogenesis by MoMuLV (LEE et al. 1981b; STORCH and CHUSED 1984). The importance of the CBA/J immune defect in oncogenesis, however, has been questioned because it has been shown that the CBA/N and CBA/J mice exhibit two additional important differences. Thus, CBA/N mice have an additional ectotropic provirus and they carry the *Rmf^r* allele of the *Rmcf* locus, as opposed to CBA/J mice which carry the *Rmcf^s* allele (HARTLEY et al. 1983).

Fv-3: This locus affects the host's immune response to the Friend virus complex in vivo and in vitro. The resistant allele of this locus (*Fv-3^r*) is recognized by its suppression of T cell mitogenesis induced by the Friend virus complex in vitro (KUMAR et al. 1978a, b). The identity of *Fv*-3, however, has been questioned because this locus is indistinguishable from the *Fv* 2 locus in *Fv* 2 congenic mice (CHESEBRO et al. 1990).

3.2.3 Genes Affecting the Size of the Target Cell Pool

Fv-2: The *Fv*-2 locus on mouse chromosome 9 controls the development of erythroleukemia by the Friend virus complex (LILLY 1970; ODAKA 1970). Alleles conferring susceptibility (*Fv-2^s*) are dominant over alleles conferring resistance (*Fv-2^r*) to the virus (CHESEBRO et al. 1990). The *Fv*-2 locus does not affect the replication of the Friend virus complex in culture (YOOSOOK et al. 1980). Its major effect, instead, is exerted on the cycling of the burst forming units-erythroid (BFU-Es), the target cells for the virus. Thus, it has been shown that the percentage of cycling BFU-Es in *Fv-2^{rr}* mice is lower than in *Fv-2^{ss}* animals (SUZUKI and

AXELRAD 1980). It is not clear, however, whether the *Fv-2* locus exerts its effect on the BFU-Es directly or indirectly through a diffusible factor produced elsewhere (SILVER and TEICH 1981).

Passage of the Friend virus complex through *Fv-2^{rr}* mice selects for variants that bypass the *Fv-2* restriction (STEEVES et al. 1970; GEIB et al. 1987). However, these virus variants have not been characterized to date.

W and Sl: These are two independent loci whose mutations are responsible for similar pleiotropic phenotypes. Both mutations are responsible for macrocytic anemia, sterility, and defects in hair pigmentation. However, the two are distinct because *W* maps to chromosome 5 while *Sl* (*steel*) maps to chromosome 10 (SILVERS 1979). In addition, mutations in *W* are responsible for a stem cell defect, while the defect of *Sl* is exerted through the bone marrow microenvironment (SILVERS 1979). Recently it was shown that the *W* mutation affects the tyrosine kinase receptor oncogene *c-kit* on chromosome 5 (CHABOT et al. 1988) and that *Sl* codes for a factor that is the natural ligand of this receptor (WITTE 1990). Multiple mutant alleles of both loci, giving rise to defects of variable severity, have been described to date (BERNSTEIN et al. 1990). Both loci affect the susceptibility to the Friend virus complex. Resistance of the *W* and *Sl* mutants to the virus appears to be the result of a decrease in the size of the target cell pool (TEICH et al. 1982).

F (flexed-tail): This locus has been mapped to mouse chromosome 13. Mutation of this locus has been associated with a transient siderocytic anemia, which occurs during periods of erythropoietic stress, and with decreased susceptibility to Friend virus-induced erythroleukemia (TEICH et al. 1982).

Nu: This mutation in mouse chromosome 11 (DAVIDSON and RODERICK 1987) is characterized by thymic aplasia. As a result, it prevents the development of T cell lymphomas. The latter effects of this mutation are similar to those of surgical removal of the thymus (GROSS 1960).

3.2.4 Genes Affecting Primarily the Type as Opposed to the Incidence of Disease

Fv-5: This is a locus which was identified in animals inoculated with the Friend virus complex and which regulates the development of anemia as opposed to polycythemia. The mechanism of action of this locus remains unknown (SHIBUYA and MAK 1982a).

4 Insertional Mutagenesis

Studies discussed earlier revealed that the main determinant of the oncogenic potential of non-acute transforming retroviruses mapped within the proviral LTR and suggested that provirus integration may play a critical role in oncogenesis

(TSICHLIS and COFFIN 1979, 1980). This hypothesis was confirmed when it was shown that provirus insertion was responsible for the activation of the c-*myc* proto-oncogene in ALV-induced bursal lymphomas (NEEL et al. 1981; HAYWARD et al. 1981). Following this, we and others began to study tumors induced by other retroviruses to determine whether insertional mutagenesis of cellular oncogenesis was a general mechanism of retrovirus oncogenesis. In these studies, it was assumed that the genes whose mutation could be responsible for tumor induction were tumor specific. Since early findings had suggested that provirus integration into the cellular genome was random, studies were initiated to determine the specificity of provirus integration in various types of retrovirus-induced tumors. The results to date have provided ample evidence that the sites targeted by provirus insertion in tumors are not random, suggesting that indeed provirus integration contributes to tumor induction (this volume). The loci targeted nonrandomly by the provirus during oncogenesis have been collectively termed common regions or common loci of integration.

The role of these provirus insertions in oncogenesis has been confirmed in a variety of ways. Experiments addressing the specificity of provirus integration in normal virus infected cells have shown that: (a) provirus insertion occurs preferentially near active genes and in close proximity to DNase I hypersensitive sites (KING et al. 1985; ROHDEWOHLD et al. 1987; VIJAYA et al. 1986; MOOSLEHNER et al. 1990) and (b) approximately 20% of provirus insertions occur by a site specific recombination mechanism, within an average of 800 sites in genomic DNA (SHIH et al. 1988). None of these experiments, however, provided evidence for a regional specificity of provirus insertion, similar to the one observed in tumor cells. This suggests that the "apparent" specificity of provirus integration in tumors is actually due to the selection of cells carrying these integration events. Other more critical experiments addressed the effects of provirus integration in loci of common integration. These experiments have shown that provirus insertion, in all the common regions of integration analyzed in sufficient detail to date, affects the expression or structure of genes involved in the control of cell growth and differentiation (this volume).

The discovery that insertional mutagenesis plays an important role in oncogenesis led to the use of provirus tagging as a method to identify genes involved in this process. The use of this approach led to the identification of new oncogenes, revealed potential oncogene interactions during the multistep process of oncogenesis, and provided clues about certain aspects of gene regulation such as the activation of genes from a long distance (this volume).

The following sections of this review will address the mechanisms by which provirus integration affects the regulation or function of oncogenes and the properties of genes known to be the targets of insertional mutagenesis.

4.1 Effects of Provirus Insertion

4.1.1 Promoter Insertion

Provirus integration, either 5' or within a given oncogene and in the same transcriptional orientation as that gene, may give rise to oncogene RNA transcripts which start in the introduced proviral LTR promoter. Transcription may initiate in the 3' or 5' proviral LTR. Initiation of trascription in the 5' LTR is followed by splicing of the resulting readthrough viral RNA transcript joining a viral splice donor site to a splice acceptor site in the cell derived part of the message. Both mechanisms give rise to hybrid mRNA trancripts containing viral and cellular sequences. When transcription starts in the 3'LTR the integrated provirus contains 5' deletions or other mutations which interfere with its transcription (NEEL et al. 1981; ROBINSON and GAGNON 1986; GOODENOW and HAYWARD 1987). It has been hypothesized that the interruption of transcription through the 3'LTR promoter, which is caused by these mutations, is required for expression of the transcriptional activity of this promoter (CULLEN et al. 1984). This mechanism of oncogene activation has been termed promoter insertion.

Activation by promoter insertion in retrovirus induced rodent hematopoietic neoplasms has been described for the following genes: *Lck*, c-Ki-*ras*, c-Ha-*ras*, *Evi*-1, c-*myb*, *Hox*-2.4, and *Mlvi*-4.

The *Lck* gene is transcribed from two promoters approximately 10 kb apart. RNA messages initiating at the proximal promoter are called type I, while those initiating at the distal promoter are called type II (VORONOVA et al. 1987; ADLER et al. 1988). Provirus insertion near *Lck* was detected in two cell lines, LSTRA and Thy 19, derived from murine T cell lymphomas induced by MoMuLV (MARTH et al. 1985; VORONOVA et al. 1986, 1987; ADLER et al. 1988). Provirus insertion in LSTRA and Thy 19 cells occurred 962 and 584 nucleotides 5' of the proximal *Lck* promoter, respectively, and in the same transcriptional orientation as the *Lck* gene. Transcription starts in the 5' viral LTR promoter. The resulting readthrough viral transcript is spliced at the *gag* splice donor site and a splice acceptor site located 5' of the type I promoter. This generates an aberrant type II *Lck* message (VORONOVA et al. 1986, 1987; ADLER et al. 1988) which is translated into a normal protein. The net result of transcription of the *Lck* gene from the introduced viral promoter is enhanced expression.

Activation of c-Ki-*ras* by promoter insertion has been observed in the murine erythroid cell line 416B, isolated from an in vitro bone marrow culture infected with F-MuLV. In this cell line, a defective 3.5 kb F-MuLV provirus, deleted in the 5'LTR *gag* and *pol* genes, was inserted between the 5' untranslated exon and the first coding exon of c-Ki-*ras* and in the same transcriptional orientation as this gene (GEORGE et al. 1986). Transcription of c-Ki-*ras* from the viral promoter results in the enhanced expression of an aberrant RNA message which is translated into a normal protein (ELLIS et al. 1982).

The *Evi*-1 locus was identified as a common region of endogeneous ecotropic provirus integration in myeloid tumors developing spontaneously in AKXD-23 mice (MUCENSKI et al. 1986, 1988c). Subsequently, it was shown that myeloid tumors induced by MoMuLV or Cas-Br-M-MuLV may also contain a provirus in the *Evi*-1 locus. Detailed characterization of two MoMuLV-induced cell lines, NFS-78 and NFS-58 (MORISHITA et al. 1988), showed that provirus insertion occurred in the intron between the *Evi*-1 exons 1 and 2 (NFS-78) or 5' of the *Evi*-1 exon 1 (NFS-58). Two hybrid RNA messages were detected in NFS-78 cells, one of which starts at the 5' LTR and splices to the *Evi*-1 exon 2, while the other starts at the 3' LTR and runs into exon 2 without splicing. In NFS-58 cells, transcription starts at the 3' LTR and continues without splicing into exon 1.

Provirus insertion near the c-*myb* proto-oncogene has been detected in myeloid tumors induced by MoMuLV or Cas-Br-M-MuLV (SHEN-ONG et al. 1986), in plasmacytoid lymphosarcomas induced by Abelson MuLV (SHEN-ONG et al. 1984; MUSHINSKY et al. 1983) which also belong to the myeloid lineage (WEINSTEIN et al. 1987), and in myeloid cell lines derived from bone marrow cultures infected with a retrovirus construct containing v-*myc* and v-*raf* oncogenes (WEINSTEIN et al. 1987). Five Abelson MuLV-induced plasmacytoid lymphosarcomas analyzed in detail to date were shown to contain an integrated copy of the helper MoMuLV provirus 5' of the first c-*myb* exon with v-*myb* homology and in the same transcriptional orientation as the c-*myb* proton-oncogene. The resulting transcripts start at the 5' LTR and, using a cryptic splice donor site in the *gag* gene, they splice to the first v-*myb* homologous c-*myb* exon. The *gag* and *myb* sequences in these transcripts are not in frame. Therefore, if a hybrid protein is generated, it should be synthesized through a translation frameshift mechanism. Alternatively, translation of these messages should start at a cryptic translation initiation site in the *gag* gene. Since the first v-*myb* exon corresponds to the fourth exon of c-*myb*, these transcripts are expected to code for a protein truncated at its NH_2-terminal. cDNA clones isolated from the plasmacytoid lymphosarcoma ABPL2 provided evidence that, in addition to the NH_2-truncation, some of the *myb* transcripts undergo aberrant splicing, introducing an additional 121 amino acids in the region between exons 6 and 7 (ROSSON et al. 1987). It should be noted that c-*myb* may also be activated by provirus insertion at the 3' end of the gene which leads to COOH-terminal truncation of the c-*myb* protein product (see also Sect. 4.1.3).

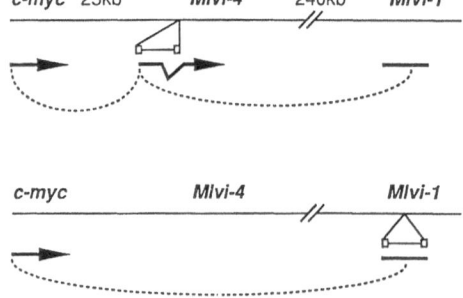

Fig. 3. Effects of provirus integration in the *Mlvi*-1 and *Mlvi*-4 loci. Activation of multiple genes some of which may be located at a long distance from the site of integration (see text for details)

The *Mlvi*-4 locus was identified as a common region of provirus insertion in T cell lymphomas induced by MoMuLV in rats (TSICHLIS et al. 1990). This locus, which maps approximately 30 kb 3' of c-*myc*, is transcribed normally only in hematopoietic organs, giving rise to 6 kb and 3 kb mRNA transcripts. The orientation of transcription of *Mlvi*-4 is the same as that of c-*myc*. Tumors containing a provirus in *Mlvi*-4 express high levels of the 3 kb and an aberrant 10 kb RNA message. Our studies to date have shown that the tumor-specific 10-kb and 3-kb messages are hybrids which are generated through a promoter insertion mechanism (Fig. 3) (TSICHLIS et al. 1990; PATRIOTIS and TSICHLIS, unpublished).

4.1.2 Enhancer Insertion

Provirus insertion near a cellular oncogene may occur at a 3' position relative to the gene and in the same transcriptional orientation. Alternatively, it may occur 5' of the gene and in the opposite transcriptional orientation. The integrated provirus is intact in most although not all cases. The net effect of these proviral insertions is the transcriptional activation of a normally silent gene (COHEN et al. 1983; GATTONI-CELLI et al. 1983; RECHAVI et al. 1982) or the transcriptional enhancement of a gene expressed normally at lower levels in the tissue of origin of the tumor (SELTEN et al. 1984). In both instances the gene is transcribed from its own promoter. If the transcription of the gene is normally under the control of more than one promoter, provirus integration may shift the balance of transcription toward one of them (REICIN et al. 1986). These phenomena, perhaps, are due to the interaction of the introduced viral enhancer with neighboring promoters. Therefore, this mechanism of transcriptional activation or enhancement has been termed enhancer insertion. Genes activated by enhancer insertion in retrovirus induced rodent neoplasms are listed in Table 4.

4.1.3 Truncation of the 5' or 3' End of the Gene: Synthesis of an Abnormal Gene Product

In some instances provirus insertion occurs within the coding sequences or within an intron between two coding exons of an oncogene. The net result of these insertions is the synthesis of mRNA transcripts coding for proteins that are truncated either at their NH_2-terminal or COOH-terminal regions. NH_2-terminal truncation of c-*myb* following provirus insertion in the 5' region of the gene has been described (see Sect. 4.1.1). COOH-terminal truncations of the c-*myb* proto-oncogene were observed in interleukin-3 (IL-3) independent myeloid leukemias induced by MoMuLV. In these tumors provirus insertion occurs in the intron between the c-*myb* exons 6 and 7 giving rise to mRNAs coding for a truncated c-*myb* protein product (SHEN-ONG et al. 1986; WEINSTEIN et al. 1986, 1987). Another gene that undergoes COOH-terminal truncation is N-*myc*. Provirus integration in the 3' end of the coding sequences of this gene in macrophage hybridomas

Table 4. Insertional mutagenesis. Effects of provirus insertion on the expression and/or structure of neighboring genes

Locus (gene)	Retrovirus	Neoplasms/cell lines	Site of Integration/ provirus orientation	Mechanism of activation	References
IL-3	IAP	Myelomonocytic leukemia WEHI-3B	5', AS	E	Ymer et al. 1985; Dührsen et al. 1990
GM-CSF	IAP	Myeloid precursor cell line DIND-1	3', S	E	Stocking et al. 1988
	Rauscher MuLV	Myeloid precursor cell line DIND-4	5', AS	E	Stocking et al. 1988
	SFFV	Myeloid precursor cell lines DIND-5 and DIND-9	5', AS	E	Stocking et al. 1988
CSF-1	Balb/c Eco	Myelomonocytic neoplasms	5', AS	E	Baumbach et al. 1987, 1988
c-fms/Fim-2	F-MuLV.	Myeloblastic leukemias	5', AS	E	Sola et al. 1988; Gisselbrecht et al. 1987; Buchberg et al. 1990
Evi-2	BXH2-Eco	Myeloid	5', AS, S	E, P[b]	Ihle et al. 1989
c-Ha-ras	MoMuLV	T cell lymphomas DA-2	5', S[a]	P[a]	George et al. 1986
c-Ki-ras	FrMuLV	Erythroid cell line 416B	5', S	P	Cuypers et al. 1984; Selten et al. 1985
Pim-1	MoMuLV	T or B cell lymphomas	3', S	E, MS	
	F-MuLV	Erythroleukemias	3', S	E, MS	Dreyfus et al. 1990
Mos	IAP	Plasmacytoma cell lines NSl, MOPC21, and derivative hybridomas	5', AS	E	Cohen et al. 1983; Gattoni-Celli et al. 1983
Lck	MoMuLV	T cell lymphoma lines LSTRA and Thy19	5', S	P	Marth et al. 1985; Voronova et al. 1986, 1987; Adler et al. 1988
c-myc	MoMuL/MCF-MμLV	T cell lymphomas	5', AS/3', S	E	Corcoran et al. 1984; Selten et al. 1984; Cuypers et al. 1984; Li et al. 1984; O'Donnel et al. 1985; Reicin et al. 1986; Steffen 1984
			Mlvi-4, S	E/Long distance	Lazo and Tsichlis 1988
			Mlvi-1/mis-1/pvt-1	Long distance	Lazo et al. 1990b; Tsichlis et al. 1989, 1990

Oncogene	Virus	Tumor/cell type	Orientation	Mechanism	References
N-myc	MoMuLV	T cell lymphomas	5', AS/3', S	E	Van Lohuizen et al. 1989a; Dolcetti et al. 1989
Myb	MLRV	Macrophage hybridomas	3', CS	E, PT	Setoguchi et al. 1989
	Cas-Br-M-MuLV	Myeloid cell line NFS 60	3', C, S	P, PT	Shen-Ong et al. 1986
	Abelson/MoMuLV	Plasmacytoid lymphosarcomas	5', C, S/3', C, S	P, E, PT	Mushinsky et al. 1983; Shen-Ong et al. 1984
Tpl-1/Ets-1	MoMuLV	T cell lymphomas	5', AS, S	E, P[b]	Bear et al. 1989
Sfpi-1/Spi-1/PU-1	SFFV	Erythroleukemias	3', S	E	Moreau-Gachelin et al. 1988; Paul et al. 1989
Evi-1	AKXD-23 Eco	Myeloid	5', S	P	Mucenski et al. 1988c
	MoMuLV	Myeloid cell line DA-1	5', S	P	Morishita et al. 1988
		Myeloid cell lines DA-3, DA-34	5', S Cb-1/Fim-3	Long distance	Bartholomew et al. 1989; Bordereaux et al. 1987
	Cas-Br-M-MuLV	Myeloid cell lines NFS 58, NFS78, NFS60	5', A, S	P, E	Morishita et al. 1988
Hox-2.4	IAP	Myelomonocytic leukemia	5', S	P	Blatt et al. 1988
p53	Abelson MuLV	Pre-B cell lymphomas	Variable	Gene inactivation/dominant negative mutations	Rotter et al. 1984
	FrMuLV	Erythroleukemias	Variable	"	Mowat et al. 1985; Rovinski et al. 1987; Chow et al. 1987; Hicks and Mowat 1988
Lyt-2	MCF MuLV	T cell lymphoma line	5', AS (35 kb)	Long distance (E)	Anson et al. 1990
Mlvi-1/Mis-1/pvt	MoMuLV	T cell lymphomas	?	?	Tsichlis et al. 1983; Tsichlis et al. 1989
Mlvi-4	MoMuLV(r)	T cell lymphomas; T cell lymphoma lines 6889, 6890	5', S?	P?	Tsichlis et al. 1990

S and AS, provirus insertion in sense or antisense orientation relative to the neighboring oncogene; 5' and 3', provirus insertion 5' or 3' of the neighboring oncogene; C integration affecting the expression of the oncogene's coding sequences, P, promoter insertion; E, enhancer insertion; PT, truncation of the oncogene product; MS, stabilization of the oncogene RNA message

[a] This provirus was rearranged, perhaps because of recombination with another provirus in the same chromosome.

[b] There is no direct experimental proof that the activation of Tpl-1/Ets-1 or Evi-2 is due to a promoter insertion mechanism. The conclusion was based only on the orientation of the proviruses integrated in these loci

removes the last six amino acids of the protein product of the gene, while it enhances its expression by an enhancer insertion mechanism (SETOGUCHI et al. 1989).

4.1.4 Stability of the RNA Message

Provirus insertion in the 3' untranslated region of a given oncogene may give rise to a truncated RNA message with normal coding capacity and altered stability.

Provirus insertion in the 3' untranslated region of pim-1 (SELTEN et al. 1985) in retrovirus-induced rodent T cell lymphomas provides an example of this phenomenon. These tumors express 2, 2.45, and 2.6 kb pim-1 mRNA transcripts. The 2 kb transcript, which is the product of early termination of transcription in the proviral LTR, lacks the sequence motif AUUUA. This motif is located in the 3' untranslated region of the gene, downstream from the site of provirus integration and has been linked to destabilization of mRNA transcripts [SHAW and KAMEN 1986). The net effect may be increased stability of the mRNA leading to a substantial increase in the steady state level of pim-1 transcripts in these tumors (CUYPERS et al. 1984; SELTEN et al. 1985). Another gene that may be activated by a similar mechanism is N-myc. Provirus insertion in the untranslated portion of the 3' exon of this gene in MCF 247-induced T cell lymphomas gives rise to truncated N-myc RNA transcripts (DOLCETTI et al. 1989).

4.1.5 Long Distance Gene Activation

In the examples discussed in the preceding paragraphs, provirus insertion affects the expression or structure of genes in its immediate vicinity. In other cases, however, provirus insertion may affect the expression of genes located at a significant distance from the site of integration. The best studied example of this phenomenon is the activation of c-myc by provirus insertion in the Mlvi-4 or Mlvi-1/mis-1/pvt-1 loci, located 30 and 270 kb 3' of c-myc, respectively (Fig. 3) (TSICHLIS et al. 1989; LAZO et al. 1990b; TSICHLIS et al. 1990). Following the observation that provirus insertion in these loci in MoMuLV-induced rat T cell lymphomas correlated with the detection of elevated steady state levels of c-myc mRNA, we constructed T cell hybrids between two rat T cell lymphomas containing a provirus in either of these loci and the murine T cell lymphoma line BW5147. These hybrids segregated the provirus containing allele from the normal allele of Mlvi-4 or Mlvi-1 and carried an intact copy of the rat c-myc. Using an S_1 nuclease protection assay we showed that the expression of the rat, but not the mouse c-myc, cosegregated with the rearranged Mlvi-4 or Mlvi-1 loci. The promoter utilization of the rat c-myc in these hybrids was not affected. These data suggest that provirus insertion may exert a long range cis effect on the expression of neighboring genes. The mechanism by which this effect is exerted has not been determined (LAZO et al. 1990b).

At least two more examples of long distance activation of oncogenes by provirus integration have been observed. These include the activation of Evi-1 by provirus insertion in the *Fim-3/Cb*-1 locus (BARTHOLOMEW et al. 1989) and the activation of the *Lyt-2* gene by provirus insertion approximately 35 kb 5' of the *Lyt*-2 promoter (D.S. ANSON, personal communication). A similar phenomenon has also been observed in mouse mammary tumor virus (MMTV)-induced mammary adenocarcinomas, where provirus insertion between the *int-2* and *Hst* genes (YOSHIDA et al. 1987) appears to be responsible for the alternate but not coordinate activation of one of these genes. In some instances the effect of provirus integration is transmitted over a relatively long distance of genomic DNA (PETERS et al. 1989).

Genetic mapping of the *Evi*-1 and *Fim-3/Cb*-1 loci, two loci of common integration in retrovirus-induced murine myeloid neoplasms, revealed that the two map to identical positions on mouse chromosome 11. However, the cloning of 110 kb of the *Fim-3/Cb*-1 locus and 80 kb of the *Evi*-1 locus revealed no overlaps (BARTHOLOMEW et al. 1989). Two myeloid cell lines, DA3 and DA34, with proviruses in the *Fim-3/Cb*-1 locus express high levels of 5 and 4 kb *Evi*-1 mRNAs. These findings indicate that provirus insertion in the *Fim-3/Cb*- 1 locus activates *Evi*-1 expression from a long distance, but they do not prove that this effect is transmitted in *cis* (BARTHOLOMEW et al. 1989).

The *Lyt*-2 (CD8) gene maps approximately 37.5 kb 3' of the *Lyt*-3 gene in the mouse genome. An *Lyt*-2$^+$ subline of the *Lyt*-2$^-$ murine T cell lymphoma line SL12.4.10, selected by immunostaining and cell sorter analysis, was found to carry a provirus integrated 35 kb 5' of *Lyt*-2 and 2.5 kb 3' of *Lyt*-3. Hybrids between the *Lyt*-2$^+$ SL12.4.10 subline and an *Lyt*-2$^-$ T cell lymphoma line express only the mutant *Lyt*-2 allele, suggesting that provirus insertion exerts a *cis* effect on the expression of the *Lyt*-2 gene (D.S. ANSON, personal communication).

4.1.6 Activation of Multiple Neighboring Genes

The finding that provirus integration may alter the transcriptional activity of genes located at a significant distance from the site of integration suggested that a single integrated provirus could affect, simultaneously, the expression of multiple neighboring genes. This was indeed demonstrated in the case of provirus insertion in the *Mlvi*-4 locus (TSICHLIS et al. 1990). Rearrangement of this locus by provirus insertion in MoMuLV-induced rat T cell lymphomas has been shown to activate c-*myc* and two additional genes, *Mlvi*-1 and *Mlvi*-4, whose expression is normally restricted to and may be developmentally regulated in T cells (Fig. 3).

4.1.7 Insertional Mutagenesis of Recessive Oncogenes: Gene Inactivation/Dominant Negative Mutations

The development of neoplasia has been associated with mutations in two types of genes: oncogenes and recessive oncogenes (antioncogens) (KNUDSON 1985). Mutations affecting oncogenes are usually heterozygous and they are associated with overexpression or enhancement of the biological activity of the gene. Conversely, mutations affecting antioncogenes are usually homozygous and associated with lack of expression of the antioncogene product. Alternatively, antioncogenes may suffer heterozygous dominant negative mutations. The mutant allele of the antioncogene in these cases produces a product which competes effectively with the product of the normal allele (LANE and BENCHIMOL et al. 1990).

It was pointed out in earlier sections of this review that provirus integration into the cellular genome is practically random and that the detection of insertion mutations affecting the same gene in multiple tumors is due to the clonal selection of randomly mutagenized cells. If we consider this observation in the context of the complexity of the mammalian genome, we can make the statistical prediction that insertion mutations in retrovirus-induced neoplasms will be exclusively heterozygous. This suggests that the great majority of insertion mutations will affect oncogenes which will be overexpressed or functionally activated by structural alterations induced by the integration of the provirus (this volume). Insertion mutations of antioncogenes will contribute to oncogenesis only if they induce dominant negative mutations or if they affect an antioncogene, one allele of which is genetically altered. Independent insertion mutations in both alleles would be expected only if each of these insertions independently contributes to the growth selection of the tumor cells.

To date only one antioncogene, *p53*, is known to be the target of insertional mutagenesis in retrovirus-induced neoplasms. Most tumors carrying insertion mutations of *p53* lack a normal *p53* allele. Mutations of *p53* have been detected in Abelson MuLV-induced pre-B cell lymphomas (WOLF and ROTTER et al. 1984; ROTTER et al. 1984) and in F-MuLV-induced erythroleukemias (MOWAT et al. 1985; ROVINSKI et al. 1987; CHOW et al. 1987; HICKS and MOWAT et al. 1988; BEN-DAVID et al. 1988). One Abelson MuLV-induced pre-B cell line (L12), which has been characterized in more detail, contains a provirus 3' of exon 1 and in the same transcriptional orientation as the *p53* gene (WOLF and ROTTER et al. 1984). The other *p53* allele could not be detected, presumably because it was deleted. Transcription of the mutant allele by promoter insertion gives rise to 3.5 and 6.5 kb sterile RNA transcripts (WOLFF and ROTTER 1984). Analysis of 31 cell lines derived from F-MuLV-induced erythroleukemias identified five which lacked expression of *p53* and two which expressed an aberrant protein (MOWAT et al. 1985). Further studies revealed that four of these cells lines carried genomic DNA rearrangements affecting both alleles of the *p53* gene. One of these cell lines was characterized in more detail and was shown to carry independent provirus insertions in both *p53* alleles. One of these insertions occurred in exon 10 while the other occurred in the region between exons 5 and 8. The proviruses involved in both insertions were defective (HICKS and MOWAT 1988). Another erythroleukemia cell line, analyzed more recently, contained an integrated spleen focus-forming provirus in the intron between exons 9 and 10. Provirus insertion in this cell line gave rise to an aberrant sterile 2.9 kb *p53* message (BEN-DAVID et al. 1988).

If we assume that the target size for provirus integration is 10 kb and that the provirus integrates randomly into the cellular DNA, then one allele of the *p53* gene will be mutated by provirus insertion in approximately 1 in 3×10^5 cells. The likelihood that the second *p53* allele will subsequently be mutated is very small (3×10^{-3} if the tumor contains 10^8 cells, 10^2 of which carry the *p53* mutation) unless the population of cells carrying one mutant *p53* allele expands selectively. Selective expansion of this cell population, however, if it occurs, would be the product of the heterozygous *p53* mutation. This is because, given the low percentage of cells carrying a provirus in one *p53* allele, the probability that they will be selected because of another independent mutation is very small. We conclude that, given the likely assumption that mutations in the two alleles of the *p53* gene occur sequentially, provirus insertions in one of them may be sufficient to give the cells a growth advantage. However, the selective advantage of cells with one mutant *p53* allele may be enhanced when the second allele is also mutated.

4.1.8 Chromosomal Rearrangements by Homologous Recombination Between Integrated Proviruses

Provirus insertion may be followed by secondary chromosomal rearrangements due to homologous recombination between integrated proviruses. One such example of a gross chromosomal rearrangement was detected in a MoMuLV-induced rat T cell lymphoma (LAZO and TSICHLIS 1988). This tumor contained an integrated provirus immediately 5' of the first exon of c-*myc* in a transcriptional orientation opposite to the gene and a second provirus at a distance of 1 ± 0.5 cm 3' of c-*myc* (N. JENKINS and N. COPELAND, unpublished). The two proviruses had integrated in opposite orientations. Homologous recombination between the 3' LTR of the c-*myc* provirus and the 5' LTR of the provirus distal to c-*myc* gave rise to an intrachromosomal inversion. One of the products of this recombination event was characterized and found to consist of an aberrant LTR structure flanked on the one side by c-*myc* and the other side by cellular sequences derived from the site of integration of the second provirus (LAZO and TSICHLIS 1988).

Another gross chromosomal rearrangement due to recombination between integrated proviruses was detected in a MoMuLV-induced murine T cell lymphoma. This tumor contained an integrated provirus in the intron between the 5' non-coding exon and the first coding exon of c-Ha-*ras* in mouse chromosome 7. Recombination between this and another provirus in the same chromosome resulted in the translocation of the c-Ha-*ras* coding sequences and enhanced expression of the gene. The translocated gene had the ability to transform fibroblasts in culture although it lacked transforming mutations in codons 12, 13, 59, and 61 (IHLE et al. 1989). A similar type of rearrangement due to homologous recombination between integrated proviruses was also detected in an ALV-induced chicken lymphoma (NOTTENBURG et al. 1987).

4.2 Genes Affected by Provirus Insertion

Provirus integration affects the expression or structure of genes involved in the transduction of proliferation or differentiation signals from the cell membrane to the nucleus. These genes, therefore, can be classified in at least four functional categories: genes coding for growth factors; growth factor receptors; membrane associated or cytoplasmic proteins involved in signal transduction; and nuclear proteins, most of which are known to have transcription factor activity. A list of these genes identified to date and their chromosomal map locations are shown in Tables 4–6. Genes whose function is unknown and loci containing genes that have not been identified to date are listed separately.

4.2.1 Growth Factors

Interleukin-3

(IL-3) is a multilineage hematopoietic growth factor which is produced by activated T cells and supports the proliferation of myeloid progenitors including those of erythrocytes, neutrophils, eosinophils, basophils, macrophages, and megakaryocytes (CLARK and KAMEN 1987). In addition it promotes the growth of pre-B and potentially pre-T cells (ITOH et al. 1990). The cellular specificity of IL-3 was suggested by the growth of numerous colonies containing cells of multiple hematopoietic lineages (LEARY et al. 1987; SIEFF et al. 1987) from IL-3 supported bone marrow cultures. These data were confirmed by studies showing that IL-3 promotes the formation of 21 day blast cell colonies from cultures of purified bone marrow progenitor cells (LEARY et al. 1987).

The cellular specificity of IL-3 for early hematopoietic progenitors suggests that it may promote the expansion of cell pools targeted by growth factors operating later in differentiation such as granulocyte/macrophage colony-stimulating factor (GM-CSF) and granulocyte colony stimulating factor (G-CSF). Expansion of these pools would in turn increase the sensitivity to these factors. Indeed it has been shown that IL-3 acts synergistically with GM-CSF, stimulating hematopoiesis in primates (DONAHUE et al. 1988). Furthermore, IL-3 appears to enhance the sensitivity to G-CSF in culture systems (LOPEZ et al. 1987).

Similar to other hematopoietic growth factors, IL-3 operates also on terminally differentiated hematopoietic cells promoting their functional activity. To date it has been shown to activate eosinophils (LOPEZ et al. 1987), stimulate mast cell growth and histamine release, and induce expression of 20α SDH and *Thy*-1 (ITOH et al. 1990).

Although IL-3 is normally secreted, this may not be necessary for it's function. To test this hypothesis the COOH-terminus of murine IL-3 was extended by adding to the 3' end of the gene an oligonucleotide coding for a four amino acid endoplasmic reticulum retention signal. Introduction of the modified IL-3 gene into hematopoietic cells promoted IL-3 independent growth despite the fact that IL-3 was not secreted. This suggested that autocrine growth may occur as a result of the intracellular action of the growth factor (DUNBAR et al. 1989). By using IL-3 dependent cell lines it was shown that IL-3 induces several phosphotyrosine-containing proteins. One of these proteins has a molecular weight of 140 kDa, is associated with the cellular membrane, and binds IL-3 with high affinity. This suggests that it may be a high affinity IL-3 receptor (ISFORT et al. 1988). This protein is not related to the cloned low affinity IL-3 receptor (ITOH et al. 1990).

Provirus insertion is responsible for the activation of the IL-3 gene in at least one myelomonocytic leukemia cell line (WEHI-3B) which is known to express IL-3 constitutively. The provirus, an intracisternal A particle (IAP), was inserted 5' and in the opposite transcriptional orientation to the IL-3 gene (YMER et al. 1985). The oncogenic potential of the IL-3 gene, suggested by these observations, was confirmed by retrovirus mediated gene transfer. Helper free virus produced from transfectal Ψ_2 cells (MANN et al. 1983) induced spleen foci in infected mice. These foci gave rise to factor independent cell lines. Transfer of these cell lines to irradiated or genetically anemic W/W^v mice (SILVERS 1979) gave rise to a myeloproliferative syndrome characterized by marked elevation of the leukocyte count, bone marrow hyperplasia, and enlargement of liver and spleen. In some animals the development of this syndrome was due to a paracrine mechanism (WONG et al. 1989).

Granulocyte-Macrophage Colony-Stimulating Factor
GM-CSF is an acidic glycoprotein which promotes the growth of neutrophil, eosinophil, and macrophage progenitors (CLARK and KAMEN 1987). In the presence of erythropoietin it also promotes the growth of erythroid precursors (SIEFF et al. 1985; LEARY et al. 1987). GM-CSF operates later than IL-3 in the process of hematopoietic cell differentiation. IL-3 therefore expands the GM-CSF target cell pool and increases the in vivo sensitivity to this factor (DONOHUE et al. 1988). Similarly to other hematopoietic growth factors, GM-CSF enhances the functional activity of the terminally differentiated cells it selects, i.e., neutrophils, ecosinophils, and macrophages (WEISBART et al. 1985; GRABSTEIN et al. 1986).

Human and murine GM-CSFs do not cross-react functionally, despite their extensive homology. Taking advantage of this observation, hybrid molecules between the human and murine genes were constructed in vitro and tested for activity in the two systems. These experiments revealed two regions, between residues 38–48 and 95–111, of the GM-CSF molecule which were critical for activity. These regions were structurally characterized by an amphiphilic helix and a disulfide bonded loop, respectively, and they were homologous in position between the human and murine growth factors. Competition assays suggested that these regions are involved in receptor binding (KAUSHANSKY et al. 1989).

The GM-CSF gene was activated by insertional mutagenesis in multiple growth factor independent variants of the GM-CSF/IL-3 dependent cell line D35. Provirus insertion occurred either 5' or 3' of the GM-CSF coding gene and activated it by an enhancer insertion mechanism (STOCKING et al. 1988).

Colony Stimulation Factor-1

CSF-1 (or M-CSF) is a T cell derived hematopoietic growth factor which supports the growth of macrophage colonies in bone marrow cultures and promotes survival and functional activity of mature macrophages (CLARK and KAMEN 1987). The mature CSF-1 protein exists in two forms: a small 40–50 kDa and a large 70–90 kDa. Both forms are disulfide linked homodimers processed from larger precursors (KAWASAKI et al. 1985; WONG et al. 1987). In humans, a small, 26 kDa, 256 amino acid precursor is coded by a 1.8 kb mRNA. A large, 61 kDa, 554 amino acid precursor, which is identical to the small precursor with an insert of 298 amino acids at position 181, is coded by a 4 kb mRNA. Processing of both precursors is associated with the removal of 32 amino acids from their NH_2-terminal domains. The additional removal of approximately 80 amino acids from the COOH-terminal domain of the small precursor gives rises to a 145 amino acid mature protein. Similarly the removal of 294 amino acids from the COOH-terminal domain of the large precursor gives rise to a 223 amino acid mature protein. The proteolytic processing of these proteins is followed by N-glycosylation and dimerization (CLARK and KAMEN 1987). The two forms of CSF-1 may differ in functional activity. Along these lines, the large form was shown to be a strong potentiator of macrophage cytotoxicity (CLARK and KAMEN 1987).

Infection of murine bone marrow cells with a c-*myc* containing retrovirus in culture and transplantation into syngeneic animals gives rise to myelomonocytic neoplasms. Analysis of one such neoplasm revealed a rearrangement of the CSF-1 gene, which was due to the integration of an endogenous ecotropic provirus 3 kb 5′ of the first CSF-1 exon and in an orientation opposite to that of the gene (BAUMBACH et al. 1987, 1988). Since cells of the myelomonocytic lineage express c-*fms* (CSF-1 receptor) (SHERR et al. 1985), the constitutive activation of CSF-1 induced by provirus insertion is responsible for the establishment of an autocrine loop leading to cell proliferation.

4.2.2 Growth Factor Receptors

c-fms/Fim-2

c-*fms* codes for the receptor of the hematopoietic growth factor CSF-1. The CSF-1 receptor is a tyrosine kinase which, on the basis of structural similarities, has been classified in the same family with the platelet-derived growth factor(PDGF) receptor and the c-*kit* proto-oncogene. Other families of tyrosine kinase/growth factor receptors include the epidermal growth factor (EGF) receptor family, the insulin receptor family, and the families of *Sea, Ret,* and *Eph* genes (RAYTER et al. 1989). Interaction of CSF-1 with c-*fms* allows progression through the G1 phase of the cell cycle, DNA synthesis, and cell division (TUSHINSKI and STANLEY 1985).

The human c-*fms* gene codes for a 972 amino acid 140 kDa glycoprotein consisting of extracellular/ligand binding, transmembrane, and intra-cytoplasmic/tyrosine kinase domains; it is expressed in cells of the monocyte/macrophage lineage (WOOLFORD et al. 1985). The mouse gene codes for a 976 amino acid protein which is 95% homologous to its human counterpart in the kinase domain and 63% in the extracellular domain (ROTHWELL and ROHRSCHNEIDER 1987). Binding of c-*fms* to its ligand (CSF-1) is followed by rapid tyrosine phosphorylation, internalization, and degradation of the protein (DOWNING et al. 1988). The viral homolog v-*fms* encodes a 180 kDa fusion glycoprotein gp180$^{gag\text{-}fms}$ which is processed to gp140$^{v\text{-}fms}$ and gp120$^{v\text{-}fms}$ (DONNER et al. 1982). The gp140$^{v\text{-}fms}$ is the mature form of the protein and it is localized on the cell membrane (MANGER et al. 1984). Glycosylation of gp140$^{v\text{-}fms}$ is critical for membrane association and transformation (NICHOLS et al. 1987). Furthermore, the membrane association, independent of its glycosylation, is necessary for v-*fms* mediated transformation (ROUSSELL et al. 1984). In contrast to the c-*fms* product gp140$^{v\text{-}fms}$ is constitutively phosphorylated on tyrosine residues. Therefore, the protein is active constitutively and independently of ligand binding (WHEELER et al. 1987). Despite this, however, gp140$^{v\text{-}fms}$ retains the ability to bind CSF-1 (SHERR 1988).

Activation of c-*fms* by provirus insertion was detected in 6 out of 42 F-MuLV-induced murine myeloid leukemias. In all cases the integrated proviruses were detected 5' of the c-*fms* proto-oncogene and they were clustered in a 2.5 kb region. The transcriptional orientation of the proviruses was opposite to that of the c-*fms* (SOLA et al. 1986, 1988). We assume that the constitutively expressed c-*fms* continues to function normally as the receptor for CSF-1 and that it transmits mitogenic signals only following binding of its normal ligand. Additional mutations, however, may transform this gene into a constitutive kinase which functions independently of its ligand. Such mutations were detected in v-*fms* and include modifications of the COOH-terminus and mutations in amino acid positions 301 and 374 (WOOLFORD et al. 1988; ROUSSEL et al. 1988).

Evi-2

This locus was identified as a common region of provirus DNA integration in BXH-2 myeloid leukemias (BUCHBERG et al. 1988). Provirus insertion activates a gene coding for a 223 amino acid product which has all the structural features of a transmembrane protein. A leucine zipper motif within the transmembrane domain of the protein suggests that *Evi*-2 may interact with other membrane components (BUCHBERG et al. 1990). Although there is no evidence that the *Evi*-2 gene product functions as a receptor, it is included here because of its structure.

The *Evi*-2 locus and the corresponding gene were mapped within an intron of the von Recklinghausen neurofibromatosis locus (BUCHBERG et al. 1990; VISCOCHIL et al. 1990; CAWTHON et al. 1990; ROBERTS 1990). The *Evi*-2 gene contains two exons and is transcribed in two 1.8 and 2.2 kb transcripts in all normal tissues. The highest level of expression was observed in macrophages, brain, and ovaries (BUCHBERG et al. 1990).

Out of 69 BXH-2 myeloid leukemias 11 contained integrated proviruses within a 14 kb region 5' of the *Evi*-2 locus. The proviruses had integrated in both transcriptional orientations. Since these tumors were not analyzed for *Evi*-2 expression, the effects of provirus integration on the transcriptional activity of this locus remain undetermined. However, it is noteworthy that three tumors contained two independent provirus integrations within the *Evi-2* locus (BUCHBERG et al. 1990). Although it was not possible to determine the clonality of these tumors, these findings raised the possibility that *Evi-2* may function as a tumor suppressor gene (see also Sects. 4.1.7 and 4.2.4).

4.2.3 Membrane Associated or Cytoplasmic Proteins Involved in Signal Transduction

c-Ha-ras and c-Ki-ras
These genes are members of a family which in mammals includes at least three closely related members: c-Ha-*ras*, c-Ki-*ras*, and N-*ras* (RAYTER et al. 1989). Several more distantly related genes have also been identified. One of these (K-*rev*-1) functions as a tumor suppressor gene (RAYTER et al. 1989; KITAYAMA et al. 1989).

All three genes code for 21 kDa proteins which are localized in the inner surface of the cell membrane (WILLINGHAM et al. 1980, 1983). The p21^{ras} proteins bind GTP and GDP with equal affinities (SCOLNICK et al. 1979; POE et al. 1985), and they have an intrinsic GTPase activity (HATTORI et al. 1985; FEUERSTEIN et al. 1987). In addition, they bind a cytoplasmic protein, GAP (GTPase activating protein), which appears to be the *ras* effector molecule and which promotes the intrinsic GTPase activity of p21^{ras} (TRAHEY and McCORMICK 1987; ADARI et al. 1988; COLEN et al. 1988; McCORMICK 1989). Since the GAP protein also interacts with tyrosine kinases, it may form a link between the *ras* and tyrosine phosphorylation pathways (ELLIS et al. 1990).

The active, signal transducing form of p21^{ras} is the GTP bound form. It is presumed that the active (GTP bound) and inactive (GDP bound) forms of p21^{ras} exist in an equilibrium which in normal cells favors the inactive form (RAYTER et al. 1989). Following an as yet unknown stimulus, the balance shifts towards the GTP bound form which interacts with GAP, the presumed p21^{ras} effector molecule, which in turn promotes the GTPase activity of p21^{ras}. The resulting hydrolysis of GTP into GDP returns the protein into its inactive form. Oncogenic mutants of *ras* genes code for proteins which frequently exhibit decreased intrinsic GTPase activity (GIBBS et al. 1984; MANNE et al. 1985; TRAHEY et al. 1987). Even the mutants with normal intrinsic GTPase activity, however, cannot be regulated by GAP (RAYTER et al. 1989). This leads to an extended interaction between the active forms of these proteins and GAP. The crystallographic characterization of p21^{ras} permitted the structural definition of its functional domains and provided rational explanations for the properties of the normal and mutant proteins (DE VOS et al. 1988; TONG et al. 1989).

The active form of p21^{ras} initiates a series of biochemical events which have been the subject of extensive investigation over the last 10 years. Based on observations that the yeast homologues of the mammalian *ras* proteins activate the adenylate cyclase system (TODA et al. 1985) and that the mammalian *ras* proteins are structurally related to the G proteins (HURLEY et al. 1984; TANABE et al. 1985), the potential role of *ras* in regulating mammalian adenylate cyclase activity was examined. These studies have provided no convincing evidence to date, however, for the activation of the mammalian adenylate cyclase system by *ras*. Other studies have shown that *ras* increases the level of diacylglycerol (DAG) without affecting the level of inositol triphosphate (IP3) (LACAL et al. 1987; FLEISCHMAN et al. 1986). Parallel studies also showed that *ras* does not affect the activity of phospholipase C (PLC) (DOWNWARD et al. 1988). The elevated level of DAG may be due to PLC independent hydrolysis of phosphatidylcholine, phosphatidylethanolamine, and phosphatidylserine (MACARA 1989), which may be secondary to the activation of the protein kinase C cascade. Activation of this cascade occurs rapidly and appears to be necessary for *ras* induced DNA synthesis (MARSHALL et al. 1989).

In addition to GTP binding and hydrolysis, the transforming activity of the *ras* proteins may also depend on their ability to bind the cell membrane. This, in turn, may be regulated by several posttranslational modifications. These include isoprenylation of cys-186, carboxymethylation, removal of the three COOH-terminal amino acids (GUTIERREZ et al. 1989; HANCOCK et al. 1989), and palmitoylation (BUSS and SEFTON 1986). The palmitoylation, in particular, is a dynamic process with a high turnover rate (MAGEE et al. 1987). The importance of these modifications in membrane association and transformation was demonstrated by mutations in cys-186. The protein products of these mutants remain cytoplasmic and they fail to transform NIH 3T3 cells (WILLUMSEN et al. 1984). On the other hand, "locking" the normal p21^{ras} on the membrane by irreversible lipid modifications leads to transformation (BUSS et al. 1989).

The function of p21^{ras} at the cellular level depends on the cell type. Thus, in NIH 3T3 cells it induces morphological transformation and DNA synthesis (STACEY and KUNG 1984). On the other hand, in PC12 pheochromocytoma cells it induces cellular differentiation (BAR-SAGI and FERAMISCO 1985). It is noteworthy that p21^{ras} may complement a variety of nuclear oncogenes in transformation (PARADA et al. 1984).

pim-1

This gene codes for a 313 amino acid cytoplasmic serine/threonine kinase which is normally expressed in fetal liver between gestation days 16 and 19 and in fetal and adult spleen and thymus (SELTEN et al. 1985, 1986; BERNS et al. 1987; DOMEN et al. 1987). The *pim*-1 gene contains six exons which extend over a 5.5 kb region and it is transcribed as a 2.8 kb RNA transcript (SELTEN et al. 1985). Activation of this gene by provirus insertion was detected in Moloney and MCF virus-induced T cell lymphomas. Provirus insertion in *pim*-1 was detected in approximately 50% of the T cell lymphomas induced within 6 months following virus inoculation. However, this frequency dropped markedly in tumors induced following a longer latency period (SELTEN et al. 1985, 1986). Occasionally, provirus insertion in *pim*-1 was also detected in murine stem cell, pre-B, and B cell neoplasms (MUCENSKI et al. 1987a).

The oncogenic potential of *pim*-1 has been tested directly in transgenic mouse experiments (see also Sect. 4.4).

Mos

This gene codes for a 37 kDa serine/threonine kinase which is localized in the cytoplasm (PAPKOFF et al. 1983) and which is expressed at very low levels in embryonic tissues and adult ovaries and testes. Even lower levels of c-mos were detected in placenta, brain, kidney, mammary gland, and epidydimis (PROPST et al. 1987).

Mos was originally identified as the oncogene transduced by the Moloney and Gazdar sarcoma viruses (TEICH 1982; TEICH et al. 1982). Both v-mos and c-mos transform NIH 3T3 cells in culture. Kinase (−) mutants of v-mos lack transforming properties suggesting that transformation depends on the kinase activity of protein (SINGH et al. 1986). Another mutation that affects the transforming activity of the mos protein product is the substitution of tyrosine at position 221 with histidine, suggesting that tyrosine phosphorylation may be important for biological activity (SINGH et al. 1988).

The expression of $p37^{c\text{-}mos}$ in the testis was localized in the round spermatids, which are haploid postmitotic germ cells (GOLDMAN et al. 1987). Expression in ovaries was localized in oocytes entering the growth phase but not in primary resting oocytes or somatic cells (KESHET et al. 1988). Subsequent experiments in frog oocytes have shown that $p37^{c\text{-}mos}$ plays a critical role in oocyte maturation and the regulation of the cell cycle (SAGATA et al. 1988, 1989). The biochemical basis of its function involves the stabilization of the maturation promoting factor (MPF), a complex between the cdc 2 kinase and cyclin (HUNT 1989). It has been proposed that the stabilization of MPF may be due to the inactivation of the cyclin protease by c-mos mediated phosphorylation (HUNT 1989). The degradation of $p37^{c\text{-}mos}$ by calpain (WATANABE et al. 1989) may contribute to the degradation of cyclin during metaphase (HUNT 1989; MURRAY et al. 1989; Murray and KIRSCHNER 1989).

Lck

The Lck gene codes for a 56 kDa src related tyrosine kinase which is expressed predominantly in Tlymphocytes (MARTH et al. 1985; VORONOVA et al. 1986). $p56^{Lck}$ is attached to the inner surface of the plasma membrane through its myristylated NH_2-terminus (HUNTER 1987).

The process of antigen dependent T cell activation depends on signals, transmitted from the T cell receptor/CD3 complex (SAIZAWA et al. 1987) and the CD4 or CD8 molecules on the surface of T lymphocytes (SAIZAWA et al. 1987; LEDBETTER et al. 1988). The CD4 and CD8 signals are generated through the interaction of these molecules with class II or class I MHC, respectively, on the membrane of antigen presenting cells (SWAIN 1983; JANEWAY JR. 1988a, b). It has been shown that signals from the antigen receptor induce rapid phosphorylation of Lck and that this process depends on two different serine kinases (VEILLETTE et al. 1988a, b). More recent studies have shown that $p56^{Lck}$ is comodulated with either CD4 or CD8 following antibody mediated cross-linking of these molecules. In addition, immune precipitation of CD4 or CD8 coprecipitates a significant fraction of the total cellular $p56^{Lck}$ protein

(VEILLETTE et al. 1988c). These findings suggest that p56^{Lck} is physically and functionally linked to CD4 or CD8 and that it may be involved in the transmission of signals required for the antigen dependent activation of T cells (VEILLETTE et al. 1988c).

4.2.4 Nuclear Proteins/Transcription Factors

Myc

The c-*myc* proto-oncogene defines a family of genes (c-*myc*, N-*myc*, R-*myc*, L-*myc*, and B-*myc*) (LE GOUY et al. 1987) which code for nuclear phosphoproteins that are expressed in a variety of cell types. The genes most commonly affected by provirus insertion in retrovirus induced neoplasms include c-*myc* and N-*myc*. The activation of c-*myc*, which is the single most commonly activated gene in retrovirus-induced tumors, occurs by multiple mechanisms. The most conspicuous of these is the long distance activation by provirus insertion 25 or 270 kb 3′ of the gene (LAZO et al. 1990b).

The transcriptional regulation of c-*myc* depends on factors that control both transcription initiation and RNA elongation. At the level of RNA elongation a partial block has been detected in exon 1 and it has been shown to play an important role in regulating the levels of mature c-*myc* RNA transcripts (BENTLEY and GROUDINE 1986; EICK et al. 1987). An additional control regulating c-*myc* function appears to operate at the level of translation. Translation of the human c-*myc* gives rise to at least two nuclear proteins with molecular weights of 64 and 67 kDa. Translation giving rise to the p67 c-*myc* protein begins in a CTG codon near the 3′ end of exon 1 (HANN et al. 1988). If p64 and p67 c-*myc* are functionally different, the synthesis of the two proteins may represent a novel mechanism for the regulation of c-*myc* function.

The deduced amino acid sequence of genes of the *myc* family revealed that they share a helix-turn-helix motif, suggesting that they may function as transcriptional regulators (MURRE et al. 1989a, b; BENEZRA et al. 1990). This suggestion was strengthened by the finding that c-*myc* may down-regulate its own expression (RABBITS et al. 1984) as well as by the finding that the various *myc* genes may negatively regulate the expression of each other (LE GOUY et al. 1987). Recent studies utilizing c-*myc*/estrogen receptor chimeras have shown that the chimeric protein affects directly or indirectly the expression of other genes in an estrogen dependent manner (EILERS et al. 1989). However, the identification of c-*myc* as a transcriptional regulator has been confounded by the lack of convincing evidence that it binds DNA in a sequence specific manner. Since c-*myc* may bind DNA as a heterodimer with another protein, this problem may be solved by the recent cloning of a gene (*Max*) which codes for a product that binds the c-*myc* protein (R. EISENMAN, personal communication). To understand the role of *myc* in oncogenesis it will be important to determine whether provirus insertions near c-*myc* occur in concert with provirus insertions near *Max*.

The codistribution of the c-*myc* product with the small nuclear ribonucleoprotein (snRNP) particles in the nucleus was interpreted in earlier studies to suggest that c-*myc* was involved in RNA processing (SPECTOR et al. 1987). No further evidence to support this idea, however, has been reported to date. More convincing were earlier studies which suggested a potential role of the c-*myc* protein in DNA replication (IGUCHI-ARIGA et al. 1987a, b; HEIKKILA et al. 1987).

At the cellular level, up-regulation of c-*myc* expression has been linked to cellular proliferation while down-regulation of c-*myc* expression has been linked to cellular differentiation (GRIEP and WESTPHAL 1988; SIMPSON et al. 1987; PHILLIPS and PARKER 1987). Exceptions to this rule, however, have been observed (LEE et al. 1987; GRAUSZ et al. 1986).

Myb

The c-*myb* gene codes for a 75 kDa-nuclear protein, $p75^{c-myb}$, while v-*myb* codes for a 45 kDa- protein, $p45^{v-myb}$. The viral protein lacks the NH_2-terminal 71 amino acids and the COOH-terminal 198 amino acids of c-*myb* (KLEMPNAUER et al. 1983a, b). Both $p45^{v-myb}$ and $p75^{c-myb}$ are localized in the nucleus and they bind DNA (KLEMPNAUER et al. 1984; MOELLING et al. 1985). DNA binding studies using $p45^{v-myb}$ expressed in bacteria revealed binding to a specific DNA sequence, pyAAC G/TG (BIEDENKAPP et al. 1988). These studies, combined with the short half-life of these proteins (BOYLE et al. 1985), suggested that *myb* may function as a transcriptional regulator.

Recent studies revealed that fusion proteins between the first 147 amino acids containing the DNA binding domain of the yeast transcriptional activator GAL4 and v-*myb* can activate transcription of the human β-globin gene if GAL4 binding sites are linked in *cis* to the β-globin gene. The v-*myb* domain found to be both necessary and sufficient for transcriptional activation was mapped between residues 204 and 254 (WESTON and BISHOP 1989). Subsequent experiments showed that the intact v-*myb* or c-*myb* products activate the β-globin gene if the *Myb* binding site was inserted upstream from the reporter gene (WESTON and BISHOP 1989). This confirmed the role of *Myb* as a transcriptional regulator.

c-*myb* is expressed primarily in immature cells of all the hematopoietic lineages (SHEINESS and GARDINIER 1984; GONDA and METCALF 1984) and secondarily in other tissues (THOMPSON et al. 1986). Differentiation of hematopoietic cells is associated with a marked drop in c-*myb* expression (THOMPSON et al. 1986). Furthermore, constitutive overexpression of c-*myb* blocks differentiation (CLARK et al. 1988) while inhibition of $p75^{c-myb}$ expression using antisense oligonucleotides inhibits cell proliferation (GEWIRTZ and CALABRETTA 1988). These findings combined suggest that c-*myb* plays a critical role in regulating hematopoietic cell differentiation and proliferation.

Tpl-1/Ets-1

Ets-1 defines a family of highly conserved genes whose products are localized in the nucleus and bind DNA (BOULUKOS et al. 1989; POGNONEC et al. 1989) through a conserved domain at their COOH-terminus (BOULUKOS et al. 1989). Characterization of the *Ets*-1 gene product revealed that *Ets*-1 codes for multiple isoforms of a nuclear phosphoprotein (KOIZUMI et al. 1990; POGNONEC et al. 1990) which contains a weak helix-turn-helix motif suggesting that it may be involved in transcriptional regulation (A. SETH, personal communication). Recent studies indeed showed that it is a transcription factor which binds a conserved sequence at position-53–34 in the LTR of MoMSV (GUNTHER et al. 1990) and the PEA3 site in the polyoma virus enhancer (WASYLYK et al. 1990). Our studies on the role of *Ets*-1 in tumor progression indicate that it promotes cellular proliferation. These data combined suggest that *Ets*-1 may be a transcription factor which activates (by a direct or indirect mechanism) genes associate with cellular proliferation. Studies on the expression and posttranslational modification of *Ets*-1 during T cell activation, however, present a challenge to this conclusion. Thus, in the early stages of T cell activation the *Ets*-1 protein product becomes transiently phosphorylated on serine and threonine residues and, as a consequence, loses the ability to bind DNA (POGNONEC et al. 1988). In addition, T cell activation is associated with an initial drop (BHAT et al. 1990) followed by a gradual increase in the steady state level of *Ets*-1 RNA spanning a 48 h period (REED et al. 1986). These data suggest that *Ets*-1 either activates genes whose function is to maintain cells in the G_o phase or it is a repressor of genes which promote cellular proliferation.

In chickens c-*Ets*-1 is transcribed from eight exons which are dispersed within a 60 kb genomic DNA region (WATSON et al. 1988). The first exon (exon 1) is absent from v-*Ets* where it is replaced by two additional exons (α and β) located upstream (WATSON et al. 1988; GEGONNE et al. 1987a; LEPRINCE et al. 1988). The differential splicing which gave rise to the transcript transduced by the virus occurs normally, although infrequently, in the spleen giving rise to an RNA transcript coding for a structurally and potentially functionally different *Ets*-1 protein product (GEGONNE et al. 1987a). In mammals this form of *Ets*-1 differential splicing has not been observed, but cDNA cloning has suggested differential splicing of exon 7 (REDDY and RAO 1988). Using an RNAse protection assay we have shown that differential splicing of exon 7 occurs in many different cell types giving rise to RNA transcripts missing exon 7 with a frequency of approximately 15%–25% (BELLACOSA and TSICHLIS, unpublished). A variant protein representing a potential translation product of the transcripts lacking exon 7 has been detected (POGNONEC et al. 1990; KOIZUMI et al. 1990).

Most of the studies on the transforming properties of *Ets*-1 have concentrated on the *gag-myb-Ets* tripartite oncogene of the acute erythroleukemia inducing avian retrovirus E26 (NUNN et al. 1984). Thus, it has been shown that the E26 virus transforms erythroblasts and to a lesser degree myeloblasts and also stimulates mitogenically chicken embryo fibroblasts in culture (JURDIE et al. 1987). These properties depend on the presence of a functional *Ets*-1 portion of the tripartite p135 transforming protein as determined by the properties of temperature sensitive mutants (HABENICHT et al. 1989; GOLAY et al. 1988). In addition, it has been shown that the *gag-myb-Ets* viral oncogene cooperates with *myc* to transform neuroretinal cells in culture (AMONGEL et al. 1989) and, when introduced into newborn mice using a Moloney based retrovirus construct, induces erythroid and myeloid leukemias (YUAN et al. 1989). Finally, the chicken c-*Ets*-1 proto-oncogene transforms NIH 3T3 cells in culture (A. SETH, personal communication).

Sfpi-1/Spi- 1/PU-1

The *Sfpi*-1/*Spi* locus was identified as a common region of provirus insertion in erythroid tumors induced by SFFV (MOREAU-GACHELIN et al. 1988; SPIRO et al. 1988; PAUL et al. 1989). Of the tumors induced by SFFV, 95% contain a provirus in this locus (MOREAU-GACHELIN et al. 1988). Provirus insertion resulted in moderate enhancement of transcription from a gene located 5′ of the site of provirus integration. cDNA cloning revealed that *Spi*-1 is identical to the *PU*-1 transcription factor (MOREAU-GACHELIN et al. 1989; PAUL et al. 1990). The *PU*-1 protein represents a novel class of transcription factors which are specific ′ for macrophages and B cells and they are members of the *Ets* oncogene family (KLEMSZ et al. 1990).

Evi-1

Provirus integration in the *Evi*-1 locus results in the activation of a cellular gene which codes for a 120 kDa-nuclear product with structural features of a zinc finger protein (MORISHITA et al. 1988). Although the precise biochemical function of this protein is not known, its cellular localization and its structure suggest that it may be a transcriptional regulator. The *Evi*-1 protein product is expressed in normal kidney tissue and in developing oocytes in the ovary, but it is not expressed in normal myeloid cells (MORISHITA et al. 1988). Therefore, its oncogenic potential in myeloid cells may be due to aberrant expression. Similar to c-*myc*, the activation of *Evi*-1 may be accomplished by proviruses that integrate at a significant distance from the *Evi*-1 gene. At least 26% of murine myeloid neoplasms induced by F-MuLV contain a provirus in a locus (*Fim*-3/*Cb*-1) which is located at least 100 kb upstream from the *Evi*-1 gene (BORDEREAUX et al. 1987). Provirus insertion in this locus activates *Evi*-1 (BARTHOLOMEW et al. 1989).

Hox-2.4

This gene is a member of the *Hox*-2 cluster of homeobox genes located on mouse chromosome 11. Insertion of an IAP particle approximately 2 kb 5′ from and in the same transcriptional orientation as the *Hox*-2.4 gene in the myeloid cell line WEH1-3B resulted in the transcriptional activation of the gene (BLATT et al. 1988). Homeobox genes code for nuclear proteins with DNA binding properties which appear to be involved in transcriptional regulation (GEHRING 1987). Expression of these genes follows a differentiation and site specific pattern in both embryos and adults of a variety of species. Isolation of homeobox gene mutants in *Drosophila* has shown that many of them are involved in pattern formation during development (GEHRING 1987). Although similar experiments have not been done in mammals, it is likely that mammalian homeobox genes are also involved in morphogenesis. Transcriptional activation of Hox-2.4, a homeobox gene normally expressed in the spinal cord, in a murine myeloid cell line contributes to the transformation phenotype, perhaps because it leads to the ectopic activation of a set of Hox-2.4 responsive genes.

p53

This gene codes for a 375 amino acid nuclear phosphoprotein which was originally detected in SV40 transformed cells because of its tight association with the SV40 large T antigen (LANE and CRAWFORD 1979; LINZER and LEVINE 1979). Subsequent studies showed that it also binds the large T antigens of other polyomaviruses (LANE and BENCHIMOL 1990) and the E1B protein of adenovirus (SARNOW et al. 1982). These associations prolong the half-life of the protein, which is normally very short (6–20 min), and allow its detection (LANE and BENCHIMOL 1990). Mutations in p53 are detected frequently in a variety of tumors. Mutant p53 protein products usually form complexes with the hsp 70 family of heat shock proteins, a phenomenon that also prolongs the half-life and enhances the steady state level of expression of the mutant proteins (STURZBECHER et al. 1987).

Tissue culture and tumor studies have provided evidence suggesting that p53 is a tumor suppressor gene or an antioncogene and that it may interact with another antioncogene, Rb. Thus, cells transformed by the adenovirus E1A protein, which inactivates the Rb gene product, lose the transformed phenotype following transfection with the normal p53 gene (FINLAY et al. 1989).

Although most of the tumor studies agree with the conclusion that p53 is an antioncogene, some of the data cannot be sufficiently interpreted on the basis of the antioncogene model. Thus, it has been shown that one Abelson MuLV-induced pre-B cell lymphoma does not express p53 and it is not oncogenic. Transfection of a mutant p53 gene in these cells renders them oncogenic (WOLF et al. 1984). The enhanced oncogenicity of p53 deficient cells following transfection of a mutant p53 gene can be interpreted only if we assume that the mutant p53 does not induce transformation only by competing with the normal gene product. Instead it may also have a direct transforming action. Microinjection of anti-p53 antibodies (MERCER et al. 1982, 1984) and introduction of antisense p53 RNA (SHOBAT et al. 1987) into normal cells prevented cell proliferation suggesting that normal p53 promotes growth. The promotion of proliferation mediated by the loss of the normal p53 function in tumor cells suggests that additional mutations may alter the cellular response to the loss of p53 in tumors.

4.2.5 Partially Characterized Genes

Mlvi-1 and Mlvi-4

Evidence discussed in Sects. 4.1.5 and 4.1.6 indicates that provirus insertions in the Mlvi-1 and Mlvi-4 loci, approximately 270 and 25 kb 3' of c-myc, respectively, activate c-*myc* by a long distance activation mechanism (LAZO and TSICHLIS 1990b). Further studies revealed that provirus insertion in these loci activates perhaps two additional genes, *Mlvi*-1 and *Mlvi*-4, which are localized in the immediate vicinity of the integrated proviruses (TSICHLIS et al. 1989; TSICHLIS et al. 1990).

Provirus insertion in the *Mlvi*-1 locus was shown to occur in three clusters separated by regions of uninterrupted DNA. The proviruses in all three clusters had integrated in a single transcriptional orientation and appeared intact. Systematic hybridization of *Mlvi*-1 clones to rat, mouse, and human genomic DNA revealed three patches of evolutionarily conserved sequences. Two of them were mapped in regions targeted by the provirus, and the third was mapped immediately 5′ to the provirus clusters. A probe derived from the conserved sequences 5′ of the integrated proviruses detected a tumor-specific RNA transcript in tumors carrying a provirus in *Mlvi*-1 or in the neighboring *Mlvi*-4 and c-*myc* loci. These findings suggest that provirus insertion in *Mlvi*-1 activates both c-*myc* and another gene which is located in the immediate vicinity of the integrated *Mlvi*-1 proviruses (TSICHLIS et al. 1989).

Provirus integration in the *Mlvi*-4 locus activates, by promoter insertion, one additional gene which maps immediately 3′ to the cluster of the *Mlvi*-4 proviruses and which is transcribed in the same orientation as c-*myc* giving rise to 3 kb and 10 kb mRNA transcripts. The *Mlvi*-4 gene is also expressed in normal thymus and spleen at very low levels, giving rise to 3 kb and 5.5 kb messages. Although *Mlvi*-4 is expressed in normal thymus, it is not expressed in MoMuLV-induced T cell lymphomas corresponding to several stages of T cell differentiation but lacking a provirus in this locus. This suggests that *Mlvi*-4 may be expressed only in a subpopulation of T cells (TSICHLIS et al. 1990).

Tpl-2 and *Gfi*-1

Tpl-2 and *Gfi*-1 are two additional genes which are only partially characterized to date (see also Sect. 4.3). Rearrangements of *Tpl*-2 occur late in oncogenesis and they are associated with tumor progression. *Tpl*-2 is expressed at very low levels in normal spleen, liver, and lung. Following provirus insertion its expression is markedly elevated (BEAR et al., unpublished).

Rearrangements of *Gfi*-1 occur during the selection of IL-2 independent variants of IL-2 dependent rat T cell lymphoma lines. *Gfi*-1 is primarily expressed in normal thymus. Provirus integration induces high levels of *Gfi*-1 expression (GILKS et al., unpublished).

4.2.6 Loci Marking Genes that Remain Undefined

A list of these loci is presented in Table 5. Their chromosomal map location is shown in Table 6.

One of these loci, *Mlvi*-2, was originally identified as a locus of common integration in MoMuLV-induced rat thymic lymphomas. Early studies showed that provirus insertions in this and the *Mlvi*-1 locus occurred in concert

Table 5. Genes activated by provirus integration in hematopoietic neoplasms induced by Type C retroviruses in rodents

Genes	References
Growth factors	
IL-3	YMER et al. 1985
GM-CSF	STOCKING et al. 1988
CSF-1	BAUMACH et al. 1988
Growth factor receptors	
c-fms/Fim-2	SOLA et al. 1986
Evi-2[a]	BUCHBERG et al. 1990
Membrane associated or cytoplasmic proteins involved in signal transduction	
c-Ha-ras	IHLE et al. 1989
c-Ki-ras	GEORGE et al. 1986
Pim-1	CUYPERS et al. 1984
Mos	COHEN et al. 1983; GATTONI-CELLI et al. 1983
Lck	VORONOVA and SEFTON 1986; MARTH et al. 1985
Nuclear proteins/transcription factors	
c-myc	CORCORAN et al. 1984; STEFFEN 1984
	SELTEN et al. 1984; LI et al. 1984
	LAZO and TSICHLIS 1988
N-myc	VAN LOHUIZEN et al. 1989a
Myb	MUSHINSKY et al. 1983, SHEN-ONG et al. 1984
	SHEN-ONG et al. 1986
Tpl-1/Ets-1	BEAR et al. 1989
Sfpi-1/Spi-1/PU-1	MOREAU-GACHELIN et al. 1988, 1989
	SPIRO et al. 1988
	PAUL et al. 1989, 1990
Evi-1	MUCENSKI et al. 1988b
Hox-2.4	BLATT et al. 1988
p53[b]	MOWAT et al. 1985
Partially characterized genes	
Mlvi-1	TSICHLIS et al. 1983; TSICHLIS et al. 1989
Mlvi-4	TSICHLIS et al. 1990
Tpl-2	BEAR et al. unpublished
Gfi-1	GILKS et al. unpublished
Other loci (gene not identified)	
Mlvi-2	TSICHLIS et al. 1985a
Mlvi-3	TSICHLIS et al. 1985a
Pim-2	BREUR et al. 1989a
Gin-1	VILLEMUR et al. 1987
Ahi-1	POIRER et al. 1988
Dsi-1	VIJAYA et al. 1987
Fim-1	SOLA et al. 1986
Fis-1	SILVER and KOZAK 1986
Fli-1	BEN-DAVID et al. 1990

[a] There is no direct experimental proof that Evi-2 codes for a receptor. Its placement under the receptor category reflects its structure.
[b] Provirus insertion is responsible for inactivation or dominant negative mutations of this gene

Table 6. Chromosomal location of common regions of provirus insertion[a]

Locus (gene)	Mouse	Human	Reference
IL-3	11	5q23-q31	BUCHBERG et al. 1988 LE BEAU et al. 1986
GM-CSF	11	5q23-q31	LE BEAU et al. 1986; HUEBNER et al. 1985
CSF-1	11	5q33.1	BUCHBERG et al. 1989; PETTENATI et al. 1987
Fim-2(fms)	18	5q33-q34	SOLA et al. 1988; HOGAN et al. 1988; LE BEAU et al. 1986
Evi-2	11	17q11	BUCHBERG et al. 1988; VISKONCHIL et al. 1990 CAWTHON et al. 1990
c-H-ras	7	11p15.5	SAKAGUCHI et al. 1984; KOZAK et al. 1983; GRZESCHIK and KAZAZIAN 1985
c-K-ras	6	6p12	SAKAGUCHI et al. 1984; GERALD and GRZESCHIK 1984
Pim-1	17	6p21	HILKENS et al. 1986; CUYPERS et al. 1986; NAGARAJAN et al. 1986
c-mos	4	8q22	CECI et al. 1989; PROPST et al. 1989; TESTA et al. 1988
Lck	4	1p35-p32	CECI et al. 1989; HUPPI et al. 1988; MARTH et al. 1986
c-myc	15	8q24	CREWS et al. 1982; TAUB et al. 1982
N-myc	12	2p24	CAMPBELL et al. 1989; KOHL et al. 1983
c-myb	10	6q22-q23	SHEN-ONG et al. 1986; HARPER et al. 1983
Tpl-1/Ets-1	9	11q23-24	BEAR et al. 1989; SACCHI et al. 1986
Sfpi-1/Spi-1/PU-1	2	11p12-11.22	MOREAU-GACHELIN et al. 1989; VAN CONG et al. 1989a
Evi-1/Fim-3/Cb-1	3	3q25-27	SOLA et al. 1988; BARTHOLOMEW et al. 1989; MUCENSKI et al. 1988; MORISHITA et al. 1990
Hox-2.4	11	17q11-22	BLATT et al. 1988; JOYNER et al. 1985
p53	11	17p13.1	BUCHBERG et al. 1988; ISOBE et al. 1986
Mlvi-1/mis-1/pvt-1	15	8q24	KOEHNE et al. 1989; KOZAK et al. 1985; TSICHLIS et al. 1985a; HENGLEIN et al. 1989
Mlvi-4	15	8q24	LAZO et al. 1990b; TSICHLIS et al. 1990
Mlvi-2	15	5p14	TSICHLIS et al. 1984; ANAGNOU et al. 1989
pim-2	17	ND	BREUER et al. 1989a
fis-1	7	11q13	SILVER and KOZAK 1986; SILVER and BUCKLER 1986; CASEY et al. 1986
Fim-1	13	6p23-p22.3	SOLA et al. 1986; VAN CONG et al. 1989b
Gin-1	19	ND	VILLEMEUR et al. 1987
Dsi-1	4	ND	VIJAYA et al. 1987
Ahi-1	10	ND	POIRIER et al. 1988

[a] Some of these loci have been mapped also in the rat genome. These include: c-myc, Mlvi-1/mis-1/pvt-1, and Mlvi-4 in chromosome 7 (SUMEGI et al. 1990; TSICHLIS et al. 1985a; TSICHLIS et al. 1989; TSICHLIS et al. 1990), Mlvi-3 in chromosome 15 (TSICHLIS et al. 1985a) and Mlvi-2 in chromosome 2 (TSICHLIS et al. 1985a), c-H-ras in chromosome 1 (SZPIRER et al. 1985) and c-K-ras in chromosome 4 (SZPIRER et al. 1985)

suggesting a potential synergistic relationship between these loci in oncogenesis (TSICHLIS et al. 1985b). Another potentially interesting observation about Mlvi-2 is its chromosomal map position. Our studies have mapped Mlvi-2 to rat chromosome 2 (TSICHLIS et al. 1985a), mouse chromosome 15 (TSICHLIS et al. 1984), and human chromosome 5p14 (ANAGNOU et al. 1989). The chromosomal mapping of Mlvi-2 in these species revealed that this locus is syntenic to the growth hormone receptor gene which has been mapped to 5p12–13 in humans and to chromosome 15 in mouse (BARTON et al. 1989). To explore the potential linkage between these two loci, the frequency of recombination between them in the genome of (*Mus musculus* x *Mus spretus*) x *Mus musculus* interspecific backcross mice was examined. Analysis of 185 backcross mice from this panel revealed no recombinants between these loci, suggesting that *Mlvi-2* maps very near the growth hormone receptor (N. JENKINS and N. COPELAND, unpublished).

4.3 Spontaneous Recurrent Provirus Integration in Virus Infected Preleukemic and Leukemic Cell Clones

Tumors are composed of cells with substantial genetic and phenotypic heterogeneity which undergo a continuous growth selection. We and others have shown that in retrovirus-induced tumors the cells selected during this process, both in vivo and in culture, contain an increasing number of integrated proviruses (BEAR et al. 1989; BREUER et al. 1989a). Given the importance of insertion mutations in retrovirus pathogenesis, these findings raised the question of whether provirus insertions in the late stages of oncogenesis were causally involved in progression. To answer this, four late provirus insertions were analyzed by testing whether they identified common loci of integration, i.e., loci targeted by the provirus in multiple tumors. Three of these late provirus insertions were selected at random while the fourth appeared during selection of IL-2 independent variants from an IL-2 dependent T cell lymphoma line. The results of these experiments were extraordinary in that they showed that all these late provirus insertions identified common loci of integration. The three loci defined by the randomly selected insertions were named *Tpl*1-3 (tumor progression loci 1-3). The fourth locus defined by provirus insertion in the cells selected for IL-2 independence was named *Gfi*-1 (growth factor independence-1 locus). The *Tpl*-1 locus is located immediately upstream of the first exon the *Ets*-1 proto-oncogene, while the *Tpl*-2 and *Gfi*-1 loci identify previously unknown genes (BEAR et al. 1989; A. MAKRIS, S. BEAR, B. GILKS, and P.N. TSICHLIS, unpublished).

These data, combined with the finding that at least 1 out of 3 of the MoMuLV-induced T cell lymphomas in rats are karyotypically normal (G. LEVAN and P. TSICHLIS, unpublished), suggest that provirus integration may be the main factor in the progression of these tumors. Furthermore, they suggest that tracing new provirus insertions in tumor cells selected for growth in vivo or in culture is a very powerful genetic tool which may allow the identification of genes contributing to the selection. This may include genes associated with metastasis or with other selectable phenotypes.

The recurrent provirus integration into the genome of virus infected preleukemic and leukemic cell clones is the source of significant genetic instability. This, coupled with the expansion of the preleukemic cell pool occurring in the early stages of retroviral oncogenesis, is reminiscent of genetic defects which predispose to the development of neoplasia either by enhancing the number of available target cells or by promoting genetic instability in somatic tissues (BELLACOSA et al. 1989). On the basis of these comparisons, it could be expected that a single virus infected individual may develope multiple independently derived tumors, a phenomenon that has been observed in rats inoculated with MoMuLV (BELLACOSA et al. 1989).

4.4 Cooperating Oncogenes

Synergism between oncogenes can be illustrated by at least two types of interactions:

1. Biochemical interactions between the products of these oncogenes: Expression of CSF-1 may enhance the growth of cells expressing the CSF-1 receptor (c-*fms*) through the establishment of an autocrine loop (BAUMBACH et al. 1987, 1988). Similarly, EGF may transform cells expressing EGF receptor although it has no effect on EGF receptor negative cells (VELU et al. 1987). The HTLV-1 *tax* gene product may enhance the expression of multiple genes by modifying the DNA binding properties of several inducible transcription factors (GREENE et al. 1989). Similarly, c-*jun* and c-*fos* are coordinately expressed and their products from heterodimers that function as transcriptional regulators (RAUSCHER et al. 1988; SASSONE-CORSI et al. 1988; GENTZ et al. 1989).
2. Interactions at the cellular level: A given oncogene may be responsible for cellular proliferation and expansion of the target cell pool which increases the chance of secondary mutations. The genes responsible for the expansion of the target cell pool cooperate indirectly with the genes affected by the secondary mutations.

Retrovirus-induced hematopoietic tumors provide significant clues about genes that may act synergistically in oncogenesis. These tumors are known to contain multiple copies of the integrated provirus in their genome. Furthermore, it is now clear that more than one of these proviruses may be involved in the process leading to the establishment of a single tumor. Studies in tumors induced by MoMuLV showed that individual tumors not only carry proviruses integrated in the domain of multiple putative oncogenes, but also that provirus insertions in some of these loci may occur in concert (TSICHLIS et al. 1985b). Similar observations were made in MMTV-induced mammary adenocarcinomas (PETERS et al. 1986). In related studies, mentioned in Sect. 4.3, retrovirus-induced neoplasms accumulate increasing numbers of integrated proviruses during progression. The genes targeted by the late provirus integrations may cooperate with other genes already active in these cells. This possibility was suggested by the distribution of provirus insertions in the *Tpl*-1 locus among 155 DNA samples from MoMuLV-induced rat T cell lymphomas. Only four provirus insertions in the *Tpl*-1 locus were detected among these samples, two of which had occurred in sublines derived from a single tumor (2772) (BEAR et al. 1989). The low frequency of provirus integration in the *Tpl*-1 locus in primary tumor tissues (1 out of 79) and cell lines (1 out of 66 DNAs tested) is in sharp contrast with the high frequency of provirus integration in this locus in cells derived from tumor 2772 (2 out of 10) ($p = 0.02$). This may mean that the effects of *Tpl*-1 activation on cell growth may depend on interactions between *Tpl*-1 and other genes activated in tumor 2772 (BEAR et al. 1989). These observations collectively suggest that certain genes, identified by provirus integration, act synergistically during oncogenesis.

More recent studies using *pim*-1 transgenic mice have confirmed and extended these observations. These mice were shown to be highly susceptible to retroviruses and chemical carcinogens (BREUER et al. 1989b, VAN LOHUIZEN et al. 1989b). Furthermore, it was shown that all the retrovirus-induced tumors in these mice carry activated c-*myc* or N-*myc* oncogenes, suggesting a synergistic relationship between *pim*-1 and genes of the *Myc* family (VAN LOHUIZEN et al. 1989b). In other studies clonal tumors induced by acute transforming retroviruses contain provirus insertions in loci of common integration that are unique to these tumors suggesting again potential synergistic interactions between *onc* genes. Thus, tumors induced by the Abelson leukemia virus contain provirus insertions in the *Ahi*-1 locus (POIRIER et al. 1988) and in vitro constructed retroviruses expressing c-*myc* induce tumors carrying provirus insertions in the c-*fms* locus (BAUMBACH et al. 1987, 1988).

Recent observations suggest that a single integrated provirus may affect coordinately the expression of multiple genes, some of which may be located at a significant distance from the site of integration. Thus, a provirus inserted in the *Mlvi*-4 locus about 30 kb 3′ of c-*myc* affects the expression of c-*myc*, *Mlvi*-4, and *Mlvi*-1 (TSICHLIS et al. 1990). It is possible that the coordinate expression of multiple neighboring genes, permits biologically significant interactions between these genes.

5 Transduction of Cellular Oncogenes

Historically, the transduction of cellular oncogenes was one of the earliest features that drew attention to retroviruses. In 1911 Rous prepared a filtered extract from a spontaneous chicken sarcoma which could transmit the disease in vivo and could morphologically alter infected culture cells. More than 60 years later it was shown that the Rous sarcoma virus contained a gene (*src*) which was dispensable for replication and which had been transduced from the cellular DNA (BISHOP and VARMUS 1985; BISHOP 1987). Following the isolation of Rous sarcoma virus a large number of retrovirus recombinants, which have transduced cellular genes, have been obtained from several species. Analysis of these recombinants led to the discovery of most of the currently known oncogenes (23).

The most likely model for the capture of cellular oncogenes suggests that the first step in this process is the integration of the provirus in or upstream of the target gene, placing it under the control of the viral promoter. The resulting hybrid transcript containing both viral and cellular sequences is packaged, replacing one of the two copies of viral RNA in the viral particles. During reverse transcription a second recombination event introduces a cDNA copy of the target oncogene into the viral sequences between the two LTR structures (BISHOP and VARMUS 1985).

The frequency of transduction of cellular proto-oncogenes varies among viruses, infected animal species, and the genetic background of individual animals within a species. Thus, feline leukemia virus in outbred cats transduces the c-*myc* proto-oncogene in a high percentage of the infected tumor bearing animals (MULLINS et al. 1983; NEIL et al. 1983). Similarly, RAV-1 induces erythroleukemia in infected 15_1 or $(K28 \times 15_1) \times K28$ chickens by transducing the c-*erb*-B proto-oncogene (FUNG et al. 1983; MILES and ROBINSON 1985).

The transduction of cellular proto-oncogenes by murine retroviruses, on the other hand, is relatively uncommon with only a handful of transforming viruses described to date. These include the Gazdar and Moloney sarcoma viruses and their variants which transduced the *Mos* proto-oncogene; the Harvey, Kirsten Rasheed, and Balb sarcoma viruses which transduced *Ras*; the FBJ and FBR murine sarcoma viruses which transduced *fos*; the Abelson leukemia virus which transduced *Abl*; the AKR thymoma virus #8 (AKT-8) (TEICH 1982; TEICH et al. 1982); the CasNS-1 acute leukemia virus which transduced the *Cbl* proto-oncogene (LANGDON et al. 1989); and the murine sarcoma virus 3611 which transduced *Raf* (RAPP et al. 1983). We recently showed that the AKT-8 virus carries a novel oncogene (*AKT*) coding for a serine threonine kinase whose catalytic domain is closely related to that of the members of the protein kinase C family (α, β, and δ) but whose regulatory domain is related to that of tyrosine kinases (A. BELLACOSA et al., in press). Of all these viruses only one, AKT-8, has been isolated from T cell lymphomas which represent by far the most frequent type of tumor induced by nontransforming retroviruses in rodents. The reason for the different frequency of transduction of cellular proto-oncogenes in different systems is unknown.

6 Transactivation of Cellular Genes by Viral Gene Products

The main factor in tumor induction by most of the animal retroviruses is the activation of cellular oncogenes by insertional mutagenesis. Alternatively, the same genes may be transduced by the replicating virus. Human and related animal retroviruses, like HTLV-I and BLV, although oncogenic, lack the ability to activate cellular oncogenes through these mechanisms. In this section we will concentrate on the molecular mechanisms of oncogenesis by HTLV-I, with the

understanding that similar mechanisms may be responsible for the pathogenicity of additional related viruses.

In addition to the genes *gag, pol,* and *env* shared by all retroviruses, the
genome of HTLV-I contains two and possibly three more overlapping genes;
(1) The transactivator X (*tax*), coding for a 353 amino acid 40 kDa nuclear protein
whose function is to up-regulate viral gene expression. The same protein may
also function as a transactivator of the HIV LTR. (2) The repressor of X (*rex*),
coding for a 189 amino acid 27 kDa nuclear protein whose function is to increase
the ratio of unspliced to spliced viral RNA messages. The same protein may exert
stimulatory or inhibitory influences on *tax* function. (3) The p21^{rex} coding for a 110
amino acid, 21 kDa nuclear protein of unknown function (Fig. 4) (LAZO and
TSICHLIS 1990).

Soon after it was isolated, HTLV-I was shown to immortalize CD4$^+$ T
lymphocytes in culture (POPOVIC et al. 1983). Although complex and not well
understood, the immortalization process appears to depend at least in part on
the function of the *tax* gene. Thus, a herpesvirus Saimiri based vector expressing
the HTLV-I *tax* gene immortalizes CD4$^+$ T cells in culture (GRASSMANN et al. 1989).
Furthermore, expression of *tax* in a variety of cell types in *tax* transgenic mice
induces cellular proliferation. Depending on the proliferating cell type a variety of
syndromes have been described in individual *tax* transgenic mouse lines,
including a neurofibromatosis-like syndrome due to proliferation of glial cells
(NERENBERG et al. 1987; HINRICHS et al. 1987), a syndrome characterized by thick
and scaly tails due to proliferating cells in the dermis and Schwann cell tumors in
the brain and tail (B. FELBER and G. PAVLAKIS, personal communication), a
Sjögren's-like syndrome due to proliferation of ductal cells in salivary and
lacrimal glands (GREEN et al. 1989a), and leukocytosis due to proliferation of
polymorphonuclear leukocytes (GREEN et al. 1989b). However, the role of *tax* in
inducing immortalization is complex as suggested by the finding that expression
of *tax* alone in culture cannot induce the immortalization phenotype.

Evidence to date suggests that the *tax* contribution to the immortalization
phenotype is mediated through the transactivation of cellular genes. The genes
known to be activated by *tax* in *trans* include *IL-2Rα* (INOUE et al. 1986; CROSS
et al. 1987; MARUYAMA et al. 1987; SIEKEVITZ et al. 1987; GREENE et al. 1986), *IL-2*
(MARUYAMA et al. 1987; SIEKEVITZ et al. 1987; GREENE et al. 1986), *GM-CSF* (NIMER
et al. 1989), and c-*fos* (NAGATA et al. 1989). The expression of these genes is upregulated in HTLV-I infected T lymphocytes. Expression of *IL-2* and *IL-2Rα* are

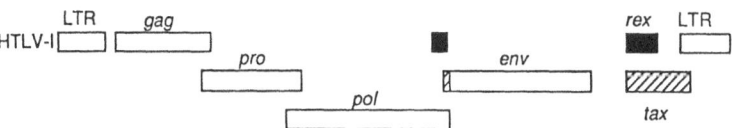

Fig. 4. Genomic organization of the HTLV-1 genome. Overlaps between *boxes* indicate overlapping
reading frames. The *tax* and *rex* mRNAs are multiply spliced. The exons contributing to the mature
message are depicted as *full bars* (*rex*) or *hatched bars* (*tax*)

responsible for establishing an autocrine loop that stimulates T cell growth (ARIMA et al. 1986; CROSS et al. 1987; MARUYAMA et al. 1987; SIEKEVITZ et al. 1987).

Secretion of GM-CSF by infected T cells may be responsible for the granulocytosis and eosinophilia observed in some patients with adult T cell leukemia (ATL) (WACHSMAN et al. 1986). Furthermore, it may contribute to the lytic bone lesions and the hypercalcemia which are observed frequently in patients with ATL (BLAYNEY et al. 1983).

The target for *tax* in the HTLV genome is positioned within the U_3 region of the LTR, between 50 and 150 bp upstream from the transcriptional start site. This target consists of three copies of a 21 bp sequence two of which are tandemly located. The third repeat is separated from the other two by a 60 bp DNA region (PASKALIS et al. 1986; FUJISAWA et al. 1986; BRADY et al. 1987). Although *tax* does not bind DNA directly, it has been shown to bind the target sequence in a complex with other proteins. The major component of this complex appears to be a 180 kDa protein (VARMUS 1988). Transactivation of the HTLV-I LTR requires the presence of a cellular transcription factor which is activated by cAMP dependent phosphorylation by protein kinase A (POTEAT et al. 1990). These findings suggest that *tax* may function as a transactivator by modifying the DNA binding properties of existing host proteins.

The effect of *tax* on the HTLV enhancer has been detected in a variety of cell types (FELBER et al. 1986; PASKALIS et al. 1986; YOSHIDA and SEIKI 1987). Nonetheless the *tax*-induced expression of the *IL-2-Rα, IL-2,* and *GM-CSF* genes is observed only in lymphocytes and it depends on the lineage and state of differentiation of the responding cells (NIMER et al. 1989; GREENE et al. 1984). This suggests that if *tax* functions by altering the DNA binding properties of cellular transcription factors, the factors involved collectively in the expression of these genes may be cell type specific. Deletion analyses of the promoter region of the *IL-2Rα* gene (Fig. 5) have implicated the NF-κb consensus sequence, − 267 to − 256 upstream from the transcriptional start site, in *tax* responsiveness (BOHNLEIN et al. 1988; BALLARD et al. 1988; LOWENTHAL et al. 1988, 1989a; LEUNG and NABEL 1988; RUBEN et al. 1988; HOYOS et al. 1989). The responsiveness of the *GM-CSF* gene to *tax* has been mapped by similar analyses at position − 53 to + 37, which is outside of the NF-κb target sequence in the promoter region of

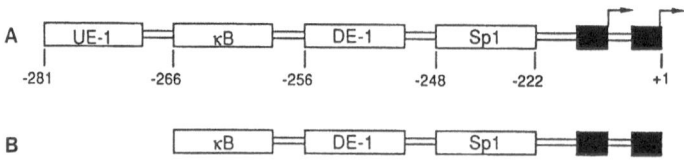

Fig. 5. Structure of the promoter and enhancer sequenences of the *IL-2Rα* gene. The two *arrows* indicate alternate transcription initiation sites. The major site is at position + 1. **A** The minimal portion of the enhancer/promoter region required for responsiveness to T cell activation stimuli. The numbers under the *boxes* define the boundaries of the marked modular enhancer elements. **B** The minimal portion of the enhancer/promoter region required for responsiveness to *tax*

this gene (NIMER et al. 1989). These findings suggest that *tax* may interact with multiple transcription factors. Alternatively, it may interact with a single cellular protein which in turn modifies multiple transcriptional regulators.

T cell activation induced by a variety of stimuli such as PHA, PMA, TNF-α, or monoclonal antibodies to the T cell receptor complex is associated with the transient expression of *IL-2 and IL- 2Rα*. On the other hand, *tax* is responsible for the constitutive expression of these genes (GOOTENBERG et al. 1981; DEPPER et al. 1984; WALDMAN et al. 1984; UCHIYAMA et al. 1985). Deletion analyses of the promoter region of the *IL-2Rα* gene have shown that the *cis* regulatory sequences required for its up-regulation during T cell activation or *tax* stimulation overlap but they are not identical (Fig. 5) (CROSS et al. 1987; BALLARD et al. 1988; LOWENTHAL et al. 1989a, b). The overlapping region required for both mitogen and *tax* responsiveness contains at least three domains: an NF-κb domain at position -266 to -256 (BALLARD et al. 1988; LOWENTHAL et al. 1989a, b; GREENE et al. 1989), a domain DE-1 (downstream element-1) which resembles the serum response element and extends from -256 to -248 (GREENE et al. 1989), and an Sp-1 binding domain extending from -244 to -239 (GREENE et al. 1989; DYNAN and TJIAN 1985).

The T cell mitogens TNF-α, and *tax* induce at least two distinct proteins that bind the NF-κb consensus sequence (BALLARD et al. 1988; LOWENTHAL et al. 1989a, b; BOHNLEIN et al. 1989; FRANZA et al. 1987; KAWAKAMI et al. 1988; LOWENTHAL et al. 1988; BOHNLEIN et al. 1988). One is the 86 kDa HIVEN86A protein while the other may be the 51 kDa NF-κb protein itself. Three additional inducible proteins (63.5 kDa, 70kDa, and 71 kDa) were also detected and they may be involved in protein–protein interactions with the 51 and 86 kDa species (GREENE et al. 1989). Induction of the NF-κb binding activity occurs even in the presence of protein synthesis inhibitors suggesting its dependence on posttranslational modifications (SEN and BALTIMORE 1986; BAEUERLE and BALTIMORE 1988). The biological significance of induction of the NF-κb binding activity was confirmed in experiments showing that the NF-κb element is sufficient to confer *tax* inducibility to the normally unresponsive promoter of the thymidine kinase gene (BOHNLEIN et al. 1988). Recent studies have shown that deletions of the DE-1 and Sp-1 domains may also affect inducibility of the *IL-2Rα* gene (GREENE et al. 1989). These observations suggest that constitutively expressed proteins that bind sites flanking the NF-κb element may be involved in the inducibility of the gene by stabilizing the interaction of the inducible proteins with DNA.

The activation of c-*fos*, a cAMP responsive gene, by *tax* is noteworthy. The potential activation of cAMP responsive genes by *tax* was originally suggested by structural and functional analyses of the HTLV-I LTR. These studies revealed a sequence, TGACG, within the 21 bp repeats which, site directed mutagenesis experiments suggested, was involved in *tax* mediated transactivation. Since this motif is related to the cAMP control element TGACGTCA (MONTMINY et al. 1986; SILVER et al. 1987), it was suggested that *tax* transactivation could be a cAMP dependent process. Recent studies, indeed, provided experimental proof for this hypothesis (POTEAT et al. 1989, 1990).

Animal non-acute retroviruses are highly oncogenic in the appropriate animal hosts. By comparison, HTLV-I is minimally oncogenic. This virus infects and immortalizes human T lymphocytes infected and maintained in culture (POPOVIC et al. 1983). However, although most infected individuals harbor infectious virus, less than 1% of them develop ATL (YOSHIDA and SEIKI 1987) or other lymphoid neoplasms (RATNER and POIESZ 1988). Furthermore, these neoplasms, which are derived from the monoclonal proliferation of CD4-positive T cells (TAKATSUKI et al. 1985), develop after an incubation period longer than 20–30 years following the initial exposure (YOSHIDA and SEIKI 1987). A potential explanation for the difference in oncogenicity between the two groups of viruses was suggested by a comparison of the biology of MoMuLV, a non-acute animal retrovirus, and HTLV-I. MoMuLV infected cells express high levels of viral RNA in vivo and produce high titers of infectious virions (TSICHLIS 1987). Conversely, freshly isolated HTLV-I infected ATL cells express undetectable levels of viral RNA and therefore produce low levels of infectious virions (KINOSHITA et al. 1989). The high levels of infectious virions produced by MoMuLV infected cells are responsible for recurrent integration of proviral DNA into the cellular genome by a mechanism that appears to involve reinfection of already infected cells (this review). Furthermore, since the MoMuLV LTR enhancer is highly active in vivo, provirus insertion influences the expression of neighboring cellular oncogenes. This is ultimately responsible for tumor induction and progression (this review). The low level of expression of HTLV-I minimizes the oncogenic potential of this virus because it decreases the frequency of provirus reintegration and the corresponding genetic instability (SEIKI et al. 1984). Moreover, the low activity of the HTLV-I LTR enhancer in vivo minimizes the effect of the integrated provirus on the expression of the adjacent cellular genes. The poor performance of HTLV-I as an insertional mutagen is indicated by the fact HTLV-I induced lymphomas do not contain the provirus integrated in common regions of integration (SEIKI et al. 1984). Thus, MoMuLV and HTLV-I differ in that MoMuLV functions both as a tumor initiator (insertional mutagen) and a tumor promoter while HTLV-I lacks initiator function.

7 Concluding Remarks

In the first part of this manuscript we reviewed the determinants of pathogenicity of the murine type C retroviruses. In our presentation we concentrated on the role of viral and host genes in determining the outcome of the retroviral infection. In the second part we have reviewed the molecular events contributing to tumor induction and progression by retroviruses. Emphasis was given primarily to insertional mutagenesis and secondarily to the transduction and transactivation of cellular oncogenes by the replicating virus.

The discovery of viral and cellular oncogenes and the finding a decade ago, that tumor induction by retroviruses was due primarily to insertion mutations in these genes, provided the intellectual framework for a revolution in cancer research. The same discoveries placed retrovirology, a formerly narrow intellectual discipline, in a central position in modern biology. Future studies are expected to provide significant insight into the function of cellular oncogenes at the molecular level and to shed new light into their role in oncogenesis and the processes of intracellular and intercellular communication. The use of insertional mutagenesis as a genetic tool to identify new oncogenes whose mutation is associated with selectable phenotypes may contribute to this understanding by uncovering novel points of gene interaction in signal transduction pathways. The study of the interactions of viral and cellular gene products, in addition to shedding new light into the mechanisms of retrovirus pathogenesis, will provide novel probes to explore the physiology of the normal cell. The study of transcriptional regulation of the proviral LTR, the study of the effects of provirus integration on the transcriptional activity of neighboring and distant genes, and the study of transactivation of genes by viral gene products are expected to continue to provide new insights into the mechanisms of regulation of gene expression in eukaryotic cells.

Future progress in these diverse areas in the field of retrovirology will expand our current understanding of biological processes and it will provide us with rational approaches towards the treatment of heoplastic and retrovirus induced diseases in humans.

Acknowledgements. We would like to thank S. Bear, J. Swantek, J.R. Testa, K. Ryder, and R. Katz for helpful discussion and numerous colleagues who have shared their data prior to publication. Special thanks to Pat Bateman for excellent secretarial assistance.

References

Adari H, Lowy DR, Williamson BM, Der C, McCormick F (1988) Guanosine triphosphate activity protein (GAP) interacts with the p21 *ras* effector binding domain. Science 240: 518–521

Adler HT, Reynolds PJ, Kelley CM, Sefton BM (1988) Transcriptional activation of *lck* by retrovirus promoter insertion between two lymphoid-specific promoters. J Virol 62: 4113–4122

Amongel P, Laudet V, Martin P, Li RP, Quatannens B, Stehelin D, Sanle S (1989) Two nuclear oncogenic proteins, p135$^{gag\text{-}myb\text{-}ets}$ and p61/63^{myc}, cooperate to induce transformation of chicken neuroretina cells. J Virol 63: 3382–3388

Anagnou NP, Economou-Pachnis A, O'Brien SJ, Modi WS, Nienhuis AW, Tsichlis PN (1989) The human homolog of the Moloney murine leukemia virus integration 2 locus maps to band p14 of chromosome 5. Genomics 5: 354–358

Arima N, Daitoku Y, Ohgaki S, Fukumori J, Tanaka H, Yamamoto Y, Fujimoto K, Onoue K (1986) Autocrine growth of interleukin-2 producing leukemic cells in a patient with adult T cell leukemia. Blood 68: 779–782

Asjö B, Skoog L, Palminger I, Wiener F, Isaak D, Cerny J, Fenyo EM (1985) Influence of genotype and the organ of origin on the subtype of T cell in Moloney lymphomas induced by transfer of preleukemic cells from athymic and thymus bearing mice. Cancer Res 45: 1040–1045

Aziz DC, Hanna Z, Jolicoeur P (1989) Severe immunodeficiency disease induced by a defective murine leukemia virus. Nature 338: 505–508

Bacon LD, Witter RL, Fadly AM (1989) Augmentation of retrovirus-induced lymphoid leukosis by Marek's disease herpesviruses in white leghorn chickens. J Virol 63: 504–512

Baeuerle PA, Baltimore D (1988) IKB: a specific inhibitor of the NF-κB transcription factor. Science 242: 540–546

Ballard DW, Böhnlein E, Lowenthal JW, Wano J, Franza BR, Greene WC (1988) HTLV-1 tax induces cellular proteins that activate the κB element in the IL-2 receptor α gene. Science 241: 1652–1655

Bar-Sagi D, Feramisco JR (1985) Microinjection of the ras oncogene protein into PC12 cells induces morphological differentiation. Cell 42: 841–848

Bartholomew C, Morishita K, Askew D, Buchberg A, Jenkins NA, Copeland NG, Ihle JN (1989) Retroviral insertions in the CB-1 / Fim-3 common site of integration activate expression of the Evi-1 gene. Oncogene 4: 529–534

Barton DE, Foellmer BE, Wood WI, Francke U (1989) Chromosome mapping of the growth hormone receptor gene in man and mouse. Cytogenet Cell Genet 50: 137–141

Bassin RH, Ruscetti S, Ali I, Haapala DK, Rein A (1982) Normal DBA/2 mouse cells synthesize a glycoprotein which interferes with MCF virus infection. Virology 123: 139–151

Baumbach WR, Stanley ER, Cole MD (1987) Induction of clonal monocyte- macrophage tumors in vivo by a mouse c-myc retrovirus: rearrangement of the CSF-1 gene as a secondary transforming event. Mol Cell Biol 7: 664–671

Baumbach WR, Colston EM, Cole MD (1988) Integration of the BALB/c ecotropic provirus into the colony-stimulating factor-1 growth factor locus in a myc retrovirus-induced murine monocyte tumor. J. Virol 62: 3151–3155

Bear SE, Tsichlis PN, Schwartz RS (1980) H-2-mediated resistance to an ecotropic type C retrovirus: localization of controlling genes and ontogenic studies of resistance. Immunogenetics 11: 451–465

Bear SE, Bellacosa A, Lazo PA, Jenkins N, Copeland N, Hanson C, Levan G, Tsichlis PN (1989) Provirus insertion in Tpl-1, an Ets-1 related oncogene, is associated with tumor progression in MoMuLV induced rat thymic lymphomas. Proc Natl Acad Sci USA 86: 7495–7499

Bellacosa A, Lazo PA, Bear SE, Shinton S, Tsichlis PN (1989) Induction of multiple independent T cell lymphomas in rats inoculated with Moloney murine leukemia. Proc Natl Acad Sci USA 86: 4269–4272

Bellacosa A, Testa JR, Staal SP, Tsichlis PN (1991) A retroviral oncogene, Akt coding for a Serine-Threonine kinase containing a src homology-2 (SH2)-like region. Science, in press

Ben-David Y, Prideaux VR, Chow V, Benchimol S, Bernstein A (1988) Inactivation of the p53 oncogene by internal deletion or retroviral integration in erythroleukemic cell lines induced by Friend leukemia viruses. Oncogene 3: 179–185

Ben-David Y, Giddens EB, Bernstein, A (1990) Identification and mapping of a common proviral integration site Fli-1 in erythroleukemia cells induced by Friend murine leukemia virus. Proc Natl Acad Sci USA 87: 1332–1336

Benezra R, Davis RL, Lockshon D, Turner DL, Weintraub H (1990) The protein Id: a negative regulator of helix-loop-helix DNA binding proteins. Cell 61: 49–59

Bentley DL, Groudine M (1986) A block to elongation is largely responsible for decreased transcription of c-myc in differentiated HL60 cells. Nature 321: 702–706

Berns A, Cuypers HT, Selten G, Domen J (1987) Pim-1 activation in T cell lymphomas. In: Kjeldgaard, Forchhammer. (eds) Proceeding Benzon Symposium on Viral Carcinogenesis. Munskgaard, Copenhagen, pp 211–224

Bernstein A, Chabot B, Dubreuil P, Reith A, Nocka K, Majumder S, Ray P, Besmer P (1990) The mouse Wlc-kit locus. Molecular control of haemopoiesis. Ciba Foundation Symposium 140. Wiley, Chichester, pp 158–172

Bhat NK, Thompson CB, Lindsten T, June CH, Fujiwara S, Koizumi S, Fisher RJ, Pappas TS (1990) Reciprocal expression of human ETS1 and ETS2 genes during T-cell activation: regulatory role for the proto-oncogene ETS1. Proc Natl Acad Sci USA 87: 3723–3727

Biedenkapp H, Borgmeyer U, Sippel AE, Klempnauer K-H (1988) Viral myb oncogene encodes a sequence-specific DNA-binding activity. Nature 335: 835–837

Bilello JA, Pitts OM, Hoffman PM (1986) Characterization of a progressive neurodegenerative disease induced by a temperature-sensitive Moloney murine leukemia virus infection. J Virol 59: 234–241

Bishop JM (1987) The molecular genetics of cancer. Science 235: 305–311

Bishop JM, Varmus H (1985) Functions and origins of retroviral transforming genes. In: Weiss R, Teich N, Varmus H, Coffin JM (eds) RNA Tumor Viruses, vol 2. Cold Spring Harbor Laboratory, New York, pp. 249–356

Blatt C, Mileham K, Haas M, Nesbitt MN, Harper ME, Simon MI (1983) Chromosomal mapping of the mink cell focus-inducing and xentropic env gene family in the mouse. Proc Natl Acad Sci USA 80: 6298–6302

Blatt C, Aberdam D, Schwartz R, Sachs L (1988) DNA rearrangement of a homeobox gene in myeloid leukemic cells. EMBO J 7: 4283–4290

Blayney DW, Jaffe ES, Fisher RI, Schechter GP, Cossman J, Robert-Guroff M, Kalyanaraman VS, Blattner WA, Gallo RC (1983) The human T-cell leukemia/lymphoma virus, lymphoma, lytic bone lesions, and hypercalcemia. Ann Intern Med 98: 144–151

Böhnlein E, Lowenthal JW, Siekevitz M, Ballard DW, Franza BR, Greene WC (1988) The same inducible nuclear proteins regulates mitogen activation of both the interleukin-2 receptor-alpha gene and type 1 HIV. Cell 53: 827–836

Böhnlein E, Ballard DW, Bogerd H, Peffer NJ, Lowenthal JW, Greene WC (1989) Induction of interleukin-2 receptor-α gene expression is regulated by post-translational activation of κB specific DNA binding proteins. J Biol Chem 264: 8475–8478

Boral AL, Okenquist SA, Lenz J (1989) Identification of the SL3-3 virus enhancer core as a T lymphoma cell-specific element. J Virol 63: 76–84

Bordereaux D, Fichelson S, Sola B, Tambourin P, Gisselbrecht S (1987) Frequent involvement of the *fim*-3 region in Friend murine leukemia virus-induced myeloblastic leukemias. J Virol 61: 4043–4045

Boulukos KE, Pognonec P, Rabault B, Begue A, Ghysdael J (1989) Definition of an *Ets*-1 protein domain required for nuclear localization in cells and DNA-binding activity in vitro. Mol Cell Biol 9: 5718–5721

Boyle WJ, Lampert MA, Li AC, Baluda MA (1985) Nuclear compartmentalization of the v-*myb* oncogene product. Mol Cell Biol 5: 3017–3023

Brady J, Jeang KT, Duvall J, Khoury G (1987) Identification of p40x-responsive regulatory sequences within the human T cell leukemia virus type I long terminal repeat. J Virol 61: 2175–2181

Breuer ML, Cuypers HT, Berns A (1989a) Evidence for the involvement of *pim*-2, a new common proviral insertion site, in progression of lymphomas. EMBO J 8: 743–747

Breuer M, Slebos R, Verbeek S et al. (1989b) Very high frequency of lymphoma induction by a chemical carcinogen in *pim*-1 transgenic mice. Nature 340: 61–63

Brightman BK, Davis BR, Fan H (1990) Preleukemic hematopoietic hyperplasia induced by Moloney murine leukemia virus is an indirect consequence of viral infection. J Virol 64: 4582–4584

Britt WJ, Chesebro B (1983) H-20 control of recovery from Friend virus leukemia: H-2D region influences the kinetics of the T lymphocyte response to Friend virus. J Exp Med 157: 1736–1745

Buchberg AM, Bedigian HG, Taylor BA, Brownell E, Ihle JN, Nagata S, Jenkins NA, Copeland NG (1988) Localization of *evi*-2 to chromosome 11: linkage to other proto-oncogene and growth factor loci using interspecific backcross mice. Oncogene Res 2: 149–165

Buchberg AM, Jenkins NA, Copeland NG (1989) Localization of the murine macrophage colony-stimulating factor gene to chromosome 3 using interspecific backcross analysis. Genomics 5: 363–367

Buchberg AM, Bedigian HG, Jenkins NA, Copeland NG (1990) *Evi*-2, a common integration site involved in murine myeloid leukemogenesis. Mol Cell Biol 10: 4658–4666

Buller RS, Ahmed A, Portis JL (1987) Identification of two forms of an endogenous murine retroviral *env* gene linked to the *Rmcf* locus. J Virol 61: 29–34

Buss JE, Sefton BM (1986) Direct identification of palmitic acid as the lipid attached to p21^{ras}. Mol Cell Biol 6: 116–122

Buss JE, Solski PA, Schaeffer JP, MacDonald MJ, Der CJ (1989) Activation of the cellular proto-oncogene product p21 *ras* by addition of a myristylation signal. Science 243: 1600–1602

Campbell GR, Zimmerman K, Blank RD, Alt FW, D'Eustachio, P (1989) Chromosomal location of N-*myc* and L-*myc* genes in the mouse. Oncogene Res 4: 47–54

Casey J, Smith R, McGillivray D, Peters G, Dickson C (1986) Characterization and chromosome assignment of the human homolog of *int*-2, a potential proto-oncogene. Mol Cell Biol 6: 502–510

Cawthon RM, Weiss R, Xu G, Viskochil D, Culver M, Stevens J, Robertson M, Dunn D, Gesteland R, O'Connell P, White R (1990) A major segment of the neurofibromatosis type 1 gene: cDNA sequence, genomic structure, and point mutations. Cell 62: 193–201

Ceci JD, Siracusa LD, Jenkins NA, Copeland NG (1989) A molecular genetic linkage map of mouse chromosome 4 including the localization of several proto-oncogenes. Genomics 5: 699–709

Chabot B, Stephenson DA, Chapman VM, Besmer P, Bernstein A (1988) The proto-oncogene c-*kit* encoding a transmembrane tyrosine kinase receptor maps to the mouse *W* locus. Nature 335: 88–89

Chatis PA, Holland CA, Hartley JW, Rowe WP, Hopkins N (1983) Role of the 3' end of the genome in determining disease specificity of Friend and Moloney murine leukemia virus. Proc Natl Acad Sci USA 80: 4408–4411

Chatis PA, Holland CA, Silver JE, Fredrickson TN, Hopkins N, Hartley JW (1984) A 3'-end fragment encompassing the transcriptional enhancer of a non-defective Friend virus confers erythro-leukemogenicity on Moloney murine leukemia virus. J Virol 52: 248–254

Chattopadhyay SK, Lander MR, Gupta S, Rands E, Lowy DR (1981) Origin of mink cytopathic focus-forming (MCF) viruses: comparison with ecotropic and xenotropic murine leukemia virus genomes. Virology 113: 465–483

Chesebro B, Wehrly K (1978) Rfv-1 and Rfv-2, two H-2 associated genes that influence recovery from Friend leukemia virus-induced splenomegaly. J Immunol 120: 1081–1085

Chesebro B, Wehrly K (1979) Identification of a non-H-2 gene (Rfv-3) influencing recovery from viremia and leukemia induced by Friend virus complex. Proc Natl Acad Sci USA 76: 425–429

Chesebro B, Wehrly K, Stimpfling J (1974) Host genetic control of recovery from Friend leukemia virus-induced splenomegaly. Mapping of a gene within the major histocompatibility complex. J Exp Med 140: 1457–1467

Chesebro B, Miyazawa M, Britt WJ (1990) Host genetic control of spontaneous and induced immunity to Friend murine retrovirus infection. Ann Rev Immunol 8: 477–499

Chow V, Ben-David Y, Bernstein A, Benchimol S, Mowat M (1987) Multistage Friend erythroleukemia: independent origin of tumor clones with normal or rearranged p53 cellular oncogenes. J Virol 61: 2777–2781

Clark SC, Kamen R (1987) The human hematopoietic colony-stimulating factors. Science 236: 1229–1237

Clark SP, Kaufhold R, Chan A, Mak TW (1985) Comparison of the transcriptional properties of the Friend and Moloney long terminal repeats: importance of tandem duplications and of the core enhancer sequence. Virology 144: 481–494

Clark MF, Kukowska-Latallo JK, Westin E, Smith M, Prochownik E (1988) Constitutive expression of a c-myb cDNA blocks Friend murine erythroleukemia cell differentiation. Mol Cell Biol 8: 884–892

Cloyd MD (1983) Characterization of target cells for MCF viruses in AKR mice. Cell 32: 217–225

Cloyd MD, Hartley JW, Rowe WP (1980) Lymphomogenicity of recombinant mink cell focus-forming murine leukemia virus. J Exp Med 151: 542–549

Coffin JM (1979) Structure, replication and recombination of retrovirus genomes: some unifying hypotheses. J Gen Virol 42: 1–26

Coffin J (1986) Genetic variation in AIDS viruses. Cell 46: 1–4

Coffin JM, Tsichlis PN, Barker CS, Voynow S, Robinson HL (1980) Variation in avian genomes. Ann NY Acad Sci USA 354: 410–425

Cohen JB, Unger T, Rechavi G, Canaani E, Givol D (1983) Rearrangement of the oncogene c-mos in mouse myeloma NSI and hybridomas. Nature 306: 797–799

Colen C, Hancock JF, Marshall CJ, Hall A (1988) The cytoplasmic protein GAP is implicated as the target for regulation by the ras gene product. Nature 332: 548–551

Corcoran LM, Adams JM, Dunn AR, Cory S (1984) Murine T cell lymphomas in which the cellular myc oncogene has been activated by retroviral insertion. Cell 37: 113–122

Crews S, Barth R, Hood L, Prehn J, Calame K (1982) Mouse c-myc oncogene is located on chromosome 15 and translocated to chromosome 12 in plasmacytomas. Science 218: 1319–1321

Cross S, Feinberg MB, Solf J, Holbrook N, Wong-Staal F, Leonard W (1987) Regulation of the human interleukin-2 receptor α chain promoter: activation of a non functional promoter by the trans-activator gene of HTLV-I. Cell 49: 47–56

Cullen BR, Lomedico PT, Ju G (1984) Transcriptional interference in avian retroviruses-implications for the promoter insertion model of leukaemogenesis. Nature 307: 241–245

Cuypers HT, Selten G, Quint W, Zijlstra M, Robanus-Maandag E, Boelens W, van Wezenbeek P, Melief C, Berns A (1984) Murine leukemia virus-induced T cell lymphomagenesis: integration of proviruses in a distinct chromosomal region. Cell 37: 141–150

Cuypers HT, Selten G, Berns A, Geurts van Kessel A (1986) Assignment of the human homolog of pim-1, a mouse gene implicated in leukemogenesis, to the pter-12 region of chromosome 6. Human Genet 72: 262–265

Davidson MT, Roderick TH (1987) Locus map of the mouse (Mus musculus). In: O'Brien SJ (ed) Genetic maps, vol 4. Cold Spring Harbor Laboratory, New York, pp 430–463

Davies MG, Kenney SC, Kamine J et al. (1987) Immediate-early gene region of human cytomegalo-virus trans-activates the promoter of human immunedeficiency virus. Proc Natl Acad Sci USA 84: 8642–8646

Davis BR, Chandy KG, Brightman BK, Gupta S, Fan H (1986) Effects of nonleukemogenic and wild-type Moloney murine leukemia virus on lymphoid cells in vivo: identification of a preleukemic shift in thymocyte subpopulations. J Virol 60: 423–430

Davis BR, Brightman BK, Chandy KG, Fan H (1987) Characterization of a preleukemic state induced by Moloney murine leukemia virus: evidence for two infection events during leukemogenesis. Proc Natl Acad Sci USA 84: 4875–4879

Debré P, Gisselbrecht S, Pozo F, Levy JP (1979) Genetic control of sensitivity to Moloney leukemia virus in mice. II. Mapping of three resistant genes within the H-2 complex. J Immunol 123: 1806–1812

Debré P, Boyer B, Gisselbrecht S, Bismuth A, Levy JP (1980) Genetic control of sensitivity to Moloney leukemia virus in mice. III. The three H-2 linked *Rmv* genes are immune response genes controlling the antiviral antibody response. Eur J Immunol 10: 914–918

Delpelchin A, Letesson JJ, Lostrie-Trussart N, Mammerickx M, Portetelle D, Burny A (1989) Bovine leukemia virus (BLV)-infected B-cells express a marker similar to the CD5 T-cell marker. Immunol Lett 20: 69–76

Depper JM, Leonard WJ, Kronke M, Waldmann TA, Green WC (1984) Augmented T cell growth factor receptor expression in HTLV-1 infected human leukemic cells. J Immunol 133: 1691–1695

DesGroseillers L, Jolicoeur P (1983) Physical mapping of the *Fv*-1 tropism host range determinant of BALB/c murine leukemia viruses. J Virol 48: 685–696

DesGroseillers L, Jolicoeur P (1984) Mapping the viral sequences conferring leukemogenicity and disease specificity in Moloney and amphotropic murine leukemia viruses. J Virol 52: 448–456

DesGroseillers L, Rassart E, Jolicoeur P (1983a) Thymotropism of murine leukemia virus is conferred by its long terminal repeat. Proc Natl Acad Sci USA 80: 4203–4207

DesGroseillers L, Villemur R, Jolicoeur P (1983b) The high leukemic potential of Gross passage A murine leukemia virus is conferred by its long terminal repeat. J Virol 47: 24–32

DesGroseillers L, Barrett M, Jolicoeur P (1984) Physical mapping of the paralysis-inducing determinant of a wild mouse ecotropic neurotropic retrovirus. J Virol 52: 356–363

DesGroseillers L, Rassart E, Robitaille Y, Jolicoeur P (1985) Retrovirus-induced spongiform encephalopathy: the 3'-end long terminal repeat containing viral sequences influence the incidence of the disease and the specificity of the neurological syndrome. Proc Natl Acad Sci USA 82: 8818–8822

De Vos AM, Tong L, Milburn MV, Matias PM, Jancarik J, Noguchi S, Nishimora S, Niora K, Ohtsuka E, Kim SH (1988) Three dimensional structure of an oncogene protein: catalytic domain of human c-H-*ras* p21. Science 239: 888–893

Dickson C, Eisenman R, Fan H (1985) Protein biosynthesis and assembly. In: Weiss R, Teich N, Varmus H, Coffin JM (eds) RNA Tumor Viruses, vol 2, 2nd edn. Cold Spring Harbor Laboratory, New York, pp 135–145

Doig D, Chesebro B (1979) Anti-Friend virus antibody is associated with recovery from viremia and loss of viral leukemia cell surface antigens in leukemic mice. Identification of *Rfv*-3 as a gene locus in influencing antibody production. J Exp Med 150: 10–19

Dolcetti R, Rizzo S, Viel A, Maestro R, DeRe V, Feriotto G, Boiocchi M (1989) N-*myc* activation by proviral insertion in MCF247-induced murine T-cell lymphomas. Oncogene 4: 1009–1014

Domen J, von Lindern J, Hermans A, Breuer M, Grosveld G, Berns A (1987) Comparison of the human and mouse *pim*-1 cDNAs: nucleotide sequence and immunological identification of the in vitro synthesized *pim*-1 protein. Oncogene Res 1: 103–112

Donahue RE, Seehra J, Metzger M, Lefebvre D, Rock B, Carbone S, Nathan DG, Garnick M, Sehgal PK, Laston D et al. (1988) Human IL-3 and GM-CSF act synergistically in stimulating hemato-poiesis in primates. Science 241: 1820–1823

Donner L, Fedele LA, Garon CF, Anderson SJ, Sherr CJ (1982) McDonough feline sarcoma virus: characterization of the molelcularly cloned provirus and its feline oncogene (v-*fms*). J Virol 41: 489–500

Dougherty JB, Temin HM (1988) Determination of the rate of base pair substitution and insertion mutations in retrovirus replication. J Virol 62: 2817–2822

Downing JR, Rettenmier CW, Sherr CJ (1988) Ligand-induced tyrosine kinase activity of the colony-stimulating factor 1 receptor in a murine macrophage cell line. Mol Cell Biol 8: 1795–1799

Downward J, DeGunzburg J, Weinberg RA (1988) p21 *ras* induced responsiveness of phospha-tidylinositol turnover to bradykinin is a receptor number effect. Proc Natl Acad Sci USA 85: 5774–5778

Dreyfus F, Sola B, Fichelson S, Varlet P, Charon M, Tambourin P, Wendling F, Gisselbrecht S (1990) Rearrangements of the *pim*-1, c-*myc*, and *p53* genes in Friend helper virus-induced mouse erythroleukemias. Leukemia 4: 590–594

Dührsen U, Stahl J, Gough NM (1990) In vivo transformation of factor-dependent hemopoietic cells: role of intracisternal A-particle transposition for growth factor gene activation. EMBO J 9: 1087–1096

Dunbar CE, Browder TM, Abrams JS, Nienhuis AW (1989) COOH-terminal-modified interleukin-3 is retained intracellularly and stimulates autocrine growth. Science 245: 1493–1496

Dynan WS, Tjian R (1985) Control of eukaryotic messenger RNA synthesis by sequence-specific DNA-binding proteins. Nature 316: 774–778

Eick D, Berger R, Polack A, Bornkamm GW (1987) Transcription of c-*myc* in human mononuclear cells is regulated by an elongation block. Oncogene 2: 61–65

Eilers M, Picard D, Yamamoto KR, Bishop JM (1989) Chimaeras of *myc* oncoprotein and steroid receptors cause hormone-dependent transformation of cells. Nature 340: 66–68

Ellis C, Moran M, McCormick F, Pawson T (1990) Phosphorylation of GAP and GAP-associated proteins by transforming and mitogenic tyrosine kinases. Nature 343: 377–381

Ellis RW, DeFeo D, Furth M, Scolnick EM (1982) Mouse cells contain two distinct *ras* gene mRNA species that can be translated into a p21 *onc* protein. Mol Cell Biol 2: 1339–1345

Evans LH, Malik FG (1987) Class II polytropic murine leukemia viruses (MuLVs) of AKR/J mice: possible role in the generation of class I oncogenic polytropic MuLVs. J Virol 61: 1882–1892

Evans LH, Morrey JD (1987) Tissue-specific replication of Friend and Moloney murine leukemia viruses in infected mice. J Virol 61: 1350–1357

Felber BK, Paskalis H, Kleinman-Ewing C, Wong-Staal F, Pavlakis GN (1986) The pX protein of HTLV-1 is a transcriptional activator of its long terminal repeats. Science 229: 675–679

Feuerstein J, Goody RS, Wittinghofer A (1987) Preparation and characterization of nucleotide-free and metal ion-free p21 apoprotein. J Biol Chem 262: 8455–8458

Finlay CA, Hinds PW, Levine AJ (1989) The p53 proto-oncogene can act as a suppressor of transformation. Cell 57: 1083–1093

Flanagan JR, Krieg AM, Max EE, Khan AS (1989) Negative control region at the 5' end of murine leukemia virus long terminal repeats. Mol Cell Biol 9: 739–746

Fleischman LF, Chahwala SB, Cantley L (1986) *Ras*-transformed cells: altered levels of phosphatidylinositol-4,5-bisphosphate and catabolites. Science 231: 407–410

Frankel WN, Stoye JP, Taylor BA, Coffin JM (1989a) Genetic analysis of endogenous xenotropic murine leukemia viruses: association with two common mouse mutations and the viral restriction locus *Fv*-1. J Virol 63: 1763–1774

Frankel WN, Stoye JP, Taylor BA, Coffin JM (1989b) Genetic identification of endogenous polytropic proviruses using recombinant inbred mice. J Virol 63: 3810–3821

Frankel WN, Stoye JP, Taylor BA, Coffin JM (1990) A linkage map of endogenous murine leukemia proviruses. Genetics 124: 221–236

Franza RB, Josephs SF, Gilman MZ, Ryan W, Clarkson B (1987) Characterization of cellular proteins recognizing the HIV enhancer using a microscale DNA-affinity precipitation assay. Nature 330: 391–395

Fujisawa J, Seiki M, Sato M, Yoshida M (1986) A transcriptional enhancer sequence of HTLV-I is responsible for *trans*-activation mediated by p40^x of HTLV-I. EMBO J 5: 713–718

Fung MC, Hapel AJ, Ymer S, Cohen DR, Johnson RM, Campbell HD, Young IG (1984) Molecular cloning of cDNA for murine interleukin-3. Nature 307: 233–237

Fung YKT, Lewis WG, Crittenden LB, Kung HJ (1983) Activation of the cellular oncogene c-*erb*-B by LTR insertion: molecular basis for induction of erythroblastosis by avian leukosis virus. Cell 33: 357–368

Gardner MB (1985) Retroviral spongiform polioencephalomyelopathy. Rev Infect Dis 7: 99–110

Gardner MB, Henderson BE, Officer JE, Rongey RW, Parker JC, Oliver C, Estes JD, Huebner RJ (1973) A spontaneous lower motor neuron disease apparently caused by indigenous type C RNA virus in wild mice. JNCI 51: 1243–1249

Gardner MB, Rasheed S, Pal BK, Estes JD, O' Brien SJ (1980) *Akvr*-1, a dominant murine leukemia virus restriction gene, is polymorphic in leukemia-prone wild mice. Proc Natl Acad Sci USA 77: 531–535

Gattoni-Celli S, Hsiao WLW, Weinstein IB (1983) Rearranged c-*mos* locus in a MOPC21 murine myeloma cell line and its persistence in hybridomas. Nature 306: 795–796

Gegonne A, Leprince D, Pognonec P, Dernis D, Raes MB, Stehelin D, Ghysdael J (1987a) The 5′ extremity of the v-*ets* oncogene of avian leukemia virus E26 encodes amino acid sequences not derived from the major c-*ets*-encoded cellular proteins. Virology 156: 177–180

Gegonne A, Leprince D, Duterque-Coquilland M, Vandenbunder B, Flaurens A, Ghysdael J, Debuire B, Stehelin D (1987b) Multiple domains for the chicken cellular sequences homologous to v-*Ets* oncogene in E26 retrovirus. Mol Cell Biol 7: 806–812

Gehring WJ (1987) Homeo boxes in the study of development. Science 236: 1245–1252

Geib RW, Anand R, Lilly F (1987) Characterization of cell lines derived from enlarged spleens induced in C57BL/6 mice by the variant BSB strain of Friend erythroleukemia virus. Virus Res 8: 61–72

Gentz R, Rauscher FJ III, Abate C, Curran T (1989) Parallel association of *fos* and *jun* leucine zippers juxtaposes DNA binding domains. Science 243: 1695–1699

George DL, Glick B, Trusko S, and Freeman N (1986) Enhanced c-Ki-*ras* expression associated with Friend virus integration in a bone marrow-derived mouse cell line. Proc Natl Acad Sci USA 83: 1651–1655

Gerald PS, Grzeschik KH (1984) Report of the committee on the genetic constitution of chromosomes 10, 11, and 12 (HGM7). Cytogenet Cell Genet 37: 103–126

Gewirtz AM, Calabretta B (1988) A c-*myb* antisense oligonucleotide inhibits normal human hematopoiesis in vitro. Science 242: 1303–1306

Gibbs JB, Sigal IS, Peo M, Scolnick EM (1984) Intrinsic GTPase activity distinguishes normal and oncogenic ras p21 molecules. Proc Natl Acad Sci USA 81: 5704–5708

Gisselbrecht S, Fichelson S, Sola B et al. (1987) Frequent c-*fms* activation by proviral insertion in mouse myeloblastic leukaemias. Nature 329: 259–261

Golay J, Introna M, Graf T (1988) A single point mutation in the v-*ets* oncogene affects both erythroid and myelomonocytic cell differentiation. Cell 55: 1147–1148

Goldman DS, Kiessling AA, Millette CF, Cooper GM (1987) Expression of c-*mos* RNA in germ cells of male and female mice. Proc Natl Acad Sci USA 84: 4509–4513

Golemis E, Li Y, Fredrickson TN, Hartley JW, Hopkins N (1989) Distinct segments within the enhancer region collaborate to specify the type of leukemia induced by nondefective Friend and Moloney viruses. J Virol 63: 328–337

Golemis EA, Speck NA, Hopkins N (1990) Alignment of U3 region sequences of mammalian type C viruses: identification of highly conserved motifs and implications for enhancer design. J Virol 64: 534–542

Gonda TJ, Metcalf D (1984) Expression of *myb*, *myc* and *fos* proto-oncogenes during the differentiation of a murine myeloid leukaemia. Nature 310: 249–251

Goodenow MM, Hayward WS (1987) 5′ long terminal repeats of *myc*-associated proviruses appear structurally intact but are functionally impaired in tumors induced by avian leukosis viruses. J Virol 61: 2489–2498

Gootenberg JE, Ruscetti FW, Mier JE, Gazdar A, Gallo RC (1981) Human cutaneous T cell lymphoma and leukemia cell lines produce and respond to T-cell growth factor. J Exp Med 154: 1403–1417

Grabstein KH, Urdal DL, Tushinski RJ, Mochizuki DY, Price VL, Cantrell MA, Gillis S, Conlong PJ (1986) Induction of macrophage tumoricidal activity by granulocyte-macrophage colony-stimulating factor. Science 232: 506–508

Grassmann R, Dengler C, Müller-Fleckenstein I, Fleckenstein B, McGuire K, Dokhelar MC, Sodroski JG, Haseltine WA (1989) Transformation to continuous growth of primary human T lymphocytes by human T-cell leukemia virus type IX-region genes transduced by a *Herpesvirus saimiri* vector. Proc Natl Acad Sci USA 86: 3351–3355

Grausz JD, Fradelizi D, Dautry F, Monier R, Lehn P (1986) Modulation of c-*fos* and c-*myc* mRNA levels in normal human lymphocytes by calcium ionophore A23187 and phorbol ester. Eur J Immunol 16: 1217–1221

Graves BJ, Eisenman RN, McKnight SL (1985a) Delineation of transcriptional control signals within the Moloney murine sarcoma virus long terminal repeat. Mol Cell Biol 5: 1948–1958

Graves BJ, Eisenberg SP, Coen DM, McKnight SL (1985b) Alternate utilization of two regulatory domains within the Moloney murine sarcoma virus long terminal repeat. Mol Cell Biol 5: 1959–1968

Green JE, Hinrichs SH, Vogel J, Jay G (1989a) Exocrinopathy resembling Sjögren's syndrome in HTLV-1 *tax* transgenic mice. Nature 341: 72–74

Green JE, Begley CG, Wagner DK, Waldmann TA, Jay G (1989b) *Trans* activation of granulocyte-macrophage colony-stimulating factor and the interleukin-2 receptor in transgenic mice carrying the human T-lymphotropic virus type I *tax* gene. Mol Cell Biol 9: 4731–4737

Green N, Hiai H, Elder JH, Schwartz RS, Khiroya RM, Thomas CY, Tsichlis PN, Coffin JM (1980) Expression of leukemogenic recombinant viruses associated with a recessive gene in HRS/J mice. J Exp Med 152: 249–264

Green WR (1983) Cell surface expression of cytotoxic T lymphocyte-defined, AKR/Gross leukemia virus-associated tumor antigens by normal AKR.H-2^b splenic B cells. J Immunol 131: 3078–3084

Greene WC, Robb RJ, Depper JM, Leonard WJ, Drogula C, Svetlik PB, Wong-Staal F, Gallo RC, Waldmann T (1984) Phorbol diester induces expression of Tac antigen on human acute T-lymphocytic leukemic cells. J Immunol 133: 1042–1047

Greene WC, Leonard WJ, Wane Y, Svetlik PB, Peffer NJ, Sodrowski JG, Rosen CA, Goh WC, Haseltine WA (1986) Trans-activator genes of HTLV-II induces IL-2 receptor and IL-2 cellular gene expression. Science 232: 877–880

Greene WC, Böhnlein E, Ballard DW (1989) HIV-1, HTLV-1 and normal T cell growth: transcriptional strategies and surprises. Immunology Today 10: 272–278

Griep AE, Westphal H (1988) Antisense myc sequences induce differentiation of F9 cells. Proc Natl Acad Sci USA 85: 6806–6810

Grodzicker T (1980) Adenovirus-SV40 hybrids in DNA tumor viruses. In: Tooze J (ed) DNA tumor viruses. Cold Spring Harbor Laboratory, New York, pp 577–614

Gross L (1960) Development of myeloid (chloro) leukemia in thymectomized C3H mice following inoculation of lymphatic leukemia virus. Proc Soc Exp Biol Med 103: 509–514

Grzeschik KH, Kazazian HH (1985) Report of the committee on the genetic constitution of chromosomes 10, 11 and 12. Cytogenet Cell Genet 40: 179–203

Gunther CV, Nye JA, Bryner RS, Graves BJ (1990) Sequence-specific DNA binding of the proto-oncoprotein ets-1 defines a transcriptional activator sequence within the long terminal repeat of the Moloney murine sarcoma virus. Genes Dev 4: 667–679

Gutierrez L, Magee AI, Marshall CJ, Hancock JF (1989) Posttranslational processing of p21^{ras} is two-step and involves carboxylmethylation and carboxy-terminal proteolysis. EMBO J 8: 1093–1098

Habenicht AJ, Goerig M, Rothe DE, Specht E, Ziegler R, Glomset JA, Graft T (1989) Early reversible reduction of leukotriene synthesis in chicken myelomonocytic transformed by a temperature-sensitive mutant of avian leukemia virus E26. Proc Natl Acad Sci USA 86: 921–924

Hancock JF, Childs J, Magee A, Marshall CJ (1989) All p21ras proteins are isoprenylated but only some are palmitoylated. Cell 57: 1167–1177

Hann SR, King MW, Bentley DL, Anderson CW, Eisenman RN (1988) A non-AUG translational initiation of c-myc exon 1 generates an N-terminally distinct protein whose synthesis is disrupted in Burkitt's lymphomas. Cell 52: 185–195

Harper ME, Franchini G, Love J, Simon MI, Gallo RC, Wong-Stahl F (1983) Chromosomal sublocalization of human c-myb and c-fes cellular onc genes. Nature 304: 169–171

Hartley JW, Wolford NK, Old LJ, Rowe, WP (1977) A new class of murine leukemia virus associated with development of spontaneous lymphomas. Proc Natl Acad Sci USA 74: 789–792

Hartley JW, Yetter RA, Morse HC III (1983) A mouse gene on chromosome 5 that restricts infectivity of mink cell focus-forming recombinant murine leukemia viruses. J Exp Med 158: 16–24

Hattori S, Ulsh LS, Halliday K, Shih TY (1985) Biochemical properties of a highly purified v-ras H protein overproduced in Escherichia coli and inhibition of its activities by a monoclonal antibody. Mol Cell Biol 5: 1449–1455

Hayward WS, Neel BG, Astrin SM (1981) Activation of a cellular onc gene by promotor insertion in ALV-Induced lymphoid leukosis. Nature 290: 475–480

Heikkila R, Schwab G, Wickstrom E, Loke SL, Pluznik DH, Watt R, Neckers LM (1987) A c-myc antisense oligodeoxynucleotide inhibits entry into S phase but not progress from G_0 to G_1. Nature 328: 445–449

Henglein B, Synovzik H, Groitl P, Bornkamn GW, Hartl P, Lipp M (1989) Three breakpoints of variant t(2;8)translocations in Burkitt's lymphoma cells fall within a region 140 kilobases distal from c-myc. Mol Cell Biol 9: 2105–2113

Hiai H, Morrissey P, Khiroya R, Schwartz RS (1977) Selective expression of xenotropic virus in congenic HRS/J (hairless) mice. Nature 270: 247–249

Hicks GG, Mowat M (1988) Integration of Friend murine leukemia virus into both alleles of the p53 oncogene in an erythroleukemic cell line. J Virol 62: 4752–4755

Hilkens J, Cuypers HT, Selten G, Kroezen V, Hilgers J, Berns A (1986) Genetic mapping of pim-1 putative oncogene to mouse chromosome 17. Somat Cell Mol Genet 12: 81–88

Hinrichs SH, Nerenberg M, Reynold RK, Khoury G, Jay G (1987) A transgenic mouse model for human neurofibromatosis. Science 237: 1340–1343

Hogan MD, O'Neill RR, Kozak CA (1986) Non-ecotropic murine leukemia viruses in BALB/c and NFS/N mice: characterization of the BALB/c *Bxv*-1 provirus and the single NFS endogenous xenotrope. J Virol 60: 980–986

Hogan MD, Halden NF, Buckler CE, Kozak CA (1988) Genetic mapping of the mouse c-*fms* proto-oncogene to chromosome 18. J Virol 62: 1055–1056

Holland CA, Hartley JW, Rowe WP, Hopkins N (1985a) At least four viral genes contribute to the leukemogenicity of murine retrovirus MCF 247 in AKR mice. J Virol 53: 158–165

Holland CA, Wozney J, Chatis PA, Hopkins N, Hartley JW (1985b) Construction of recombinants between molecular clones of murine retrovirus MCF 247 and Akv: determinant of an in vitro host range property that maps in the long terminal repeat. J Virol 53: 152–157

Holland CA, Thomas CY, Chattopadhyay SK, Koehne C, O'Donnell PV (1989) Influence of enhancer sequences on thymotropism and leukemogenicity of mink cell focus-forming viruses. J Virol 63: 1284–1292

Horvat TR, Wood C, Balachandran N (1989) Transactivation of human immunodeficiency virus promoter by human herpes virus 6. J Virol 63: 970–973

Hoyos B, Ballard DW, Böhnlein E, Siekevitz M, Greene WC (1989) Kappa B-specific DNA binding proteins: role in the regulation of human interleukin-2 gene expression. Science 244: 457–460

Huebner K, Isobe M, Croce CM, Golde DW, Kaufman SE, Gasson JC (1985) The human gene encoding GM-CSF is at 5q21-q32, the chromosome region deleted in the 5q-anomaly. Science 230: 1282–1285

Hunt T (1989) Under arrest in the cell cycle. Nature 342: 483–484

Hunter T (1987) A thousand and one protein kinases. Cell 50: 823–829

Huppi K, Mock BA, Schricker P, D'Hoostelaere LA, Potter M (1988) Organization of the distal end of mouse chromosome 4. Curr Top Microbiol Immunol 137: 276–288

Hurley JB, Simon MI, Teplow DB, Robishalu JD, Gilman AG (1984) Homologies between signal transducing G-proteins and *ras* gene products. Science 226: 860–862

Iguchi-Ariga SMM, Itani T, Yamaguchi M, Ariga H (1987a) C-*myc* protein can be substituted for SV40 T-antigen DNA replication. Nucleic Acids Res 15: 4889–4899

Iguchi-Ariga SMM, Itani T, Kiji Y, Ariga H (1987b) Possible function of the c-*myc* product: promotion of cellular DNA replication. EMBO J 6: 2365–2371

Ihle JN, Enjuanes L, Lee JC et al. (1982) The immune response to C-type viruses and its potential role in leukemogenesis. Curr Top Microbiol Immunol 101: 31–49

Ihle JN, Smith-White B, Sisson B, Parker D, Blair DG, Schultz A, Kozak C, Lunsford RD, Askew D, Weinstein Y, Isport RJ (1989) Activation of the c-H- *ras* protooncogene by retrovirus insertion and chromosomal rearrangement in a Moloney leukemia virus-induced T cell leukemia. J Virol 63: 2959–2966

Ikeda H, Sugimura H (1989) *Fv*-4 resistance gene: a truncated endogenous murine leukemia virus with ecotropic interference properties. J Virol 63: 5405–5412

Ikeda H, Laigret F, Martin MA, Repaske R (1985) Characterization of a molecularly cloned retroviral sequence associated with *Fv*-4 resistance. J Virol 55: 768–777

Inoue J, Seiki M, Taniguchi T, Tsuru S, Yoshida M (1986) Induction of interleukin-2 receptor gene expression by p40 encoded by human T-cell leukemia virus type I. EMBO J 5: 2883–2888

Isfort RJ, Stevens D, May WS, Ihle JN (1988) Interleukin-3 binds to a 140-kDa phosphotyrosine-containing cell surface protein. Proc Natl Acad Sci USA 85: 7982–7986

Ishimoto A, Hartley JW, Rowe WP (1977) Detection and quantitation of phenotypically mixed viruses: mixing of ecotropic and xenotropic murine leukemia viruses. Virology 81: 263–269

Ishimoto A, Adachi A, Sakai K, Matsuyama M (1985) Long terminal repeat of Friend-MCF virus contains the sequence responsible for erythroid leukemia. Virology 141: 30–42

Ishimoto A, Takimoto M, Adachi A, Kakuyama M, Kato S, Kakimi K, Fukuoka K, Ogiu T, Matsuyama M (1987) Sequences responsible for erythroid and lymphoid leukemia in the long terminal repeats of Friend mink cell focus-forming and Moloney murine leukemia viruses. J Virol 61: 1861–1866

Isobe M, Emanuel BS, Givol D, Oren M, Croce CM (1986) Localization of gene for human p53 tumour antigen to band 17p13. Nature 320: 84–85

Itoh N, Yonehara S, Schreurs J, Gorman DM, Maruyama K, Ishii A, Yahara I, Arai K, Miyajima A (1990) Cloning of an interleukin-3 receptor gene: a member of a distinct receptor gene family. Science 247: 324–327

Janeway CA Jr (1988a) Molecular recognition. Frontiers of the immune system. Nature 333: 804–806

Janeway CA Jr (1988b) T-cell development. Accessories or coreceptors? Nature 335: 208–210

Jenkins NA, Copeland NG, Taylor BA, Lee BK (1982) Organization, distribution and stability of endogenous ecotropic murine leukemia virus DNA in chromosomes of *Mus musculus*. J Virol 43: 26–36

Johnson PF, Lanschulz WH, Graves BJ, McKnight SL (1987) Identification of a rat liver nuclear protein that binds to the enhancer core element of three animal viruses. Genes 1: 133–146

Jolicoeur P, Rassart E (1981) Fate of unintegrated viral DNA in *Fv*-1 permissive and resistant mouse cells infected with murine leukemia virus. J Virol 37: 609–619

Joyner AL, Lebo RV, Kan YW, Tjian R, Cox DR, Martin GR (1985) Comparative chromosome mapping of a conserved homeo box region in human and mouse (Abstract). Cytogenet Cell Genet 40: 663

Jurdie P, Benchaibi M, Gandrillon O, Samarut J (1987) Transforming and mitogenic effects of avian leukemia virus E28 on chicken hematopoietic cells and fibroblasts, respectively, correlate with level of expression of the provirus. J Virol 61: 3058–3065

Kai K, Furuta T (1984) Isolation of paralysis-inducing murine leukemia viruses from Friend virus passaged in rats. J Virol 50: 970–973

Kaushansky K, Shoemaker SG, Alfaro S, Brown C (1989) Hematopoietic activity of granulocyte/macrophage colony-stimulating factor is dependent upon two distinct regions of the molecule: functional analysis based upon the activities of interspecies hybrid growth factors. Proc Natl Acad Sci USA 86: 1213–1217

Kawasaki ES, Ladner MB, Wang AM, Arsdell JV, Warren MK, Coyne MY, Schweickart VL, Lee MT, Wilson KJ, Boosman A, Stanley ER, Ralph P, Mark DF (1985) Molecular cloning of a complementary DNA encoding human macrophage-specific colony-stimulating factor (CSF-1). Science 230: 291–296

Kawakami K, Scheiderest C, Roeder RG (1988) Identification and purification of a human immunoglobulin-enhancer-binding protein (NF-κB) that activates transcription from a human immunodeficiency virus type 1 promoter in vitro. Proc Natl Acad Sci USA 85: 4700–4704

Keshet E, Rosenberg MP, Mercer JA, Propst F, Vande Woude GF, Jenkins NA, Copeland NG (1988) Developmental regulation of ovarian-specific *mos* expression. Oncogene 2: 235–240

Khan AS (1984) Nucleotide sequence analysis establishes the role of endogenous murine leukemia virus DNA segments in formation of recombinant mink cell focus-forming murine leukemia viruses. J Virol 50: 864–871

King DP, Jones PP (1983) Induction of Ia and H-2 antigens on a macrophage cell line by immune interferon. J Immunol 131: 315–318

King W, Patel MD, Lobel LI, Goff SP, Nguyen-Huu MC (1985) Insertion mutagenesis of embryonal carcinoma cells by retroviruses. Science 228: 554–558

Kinoshita T, Shimoyama M, Tobinai K, Ito M, Ito SI, Ikeda S, Tajima K, Shimotohno K, Sugimura T (1989) Detection of mRNA for the *tax₁/rex₁* gene of human T cell leukemia virus type I in fresh peripheral blood mononuclear cells of adult T cell leukemia patients and viral carriers by using the polymerase chain reaction. Proc Natl Acad Sci USA 86: 5620–5624

Kitayama H, Sugimoto Y, Matsuzaki T, Ikawa Y, Noda M (1989) A *ras*-related gene with transformation suppressor activity. Cell 56: 77–84

Klempnauer KH, Gonda TJ, Bishop JM (1983a) Nucleotide sequence of the retroviral leukemia gene v-*myb* and its cellular progenitor c-*myb*: the architecture of a transduced oncogene. Cell 31: 453–463

Klempnauer KH, Ramsay G, Bishop JM, Moscovici MG, Moscovici C, McGrath JP, Levinson AD (1983b) The product of the retroviral transforming gene v myb is a truncated version of the protein encoded by the cellular oncogene c-*myb*. Cell 33: 345–355

Klempnauer KH, Symonds G, Evan GI, Bishop JM (1984) Subcellular localization of proteins encoded by oncogenes of avian myeloblastosis virus and avian leukemia virus E26 and by the chicken c-*myb* gene. Cell 37: 537

Klemsz MJ, McKercher SR, Celada A, Van Bueren C, Maki RA (1990) The macrophage and B cell specific transcription factor Pu.1 is related to the *ets* oncogene. Cell 61: 113–124

Knudson AG Jr (1985) Hereditary cancer, oncogenes, and antioncogenes. Cancer Res 45: 1437–1443

Koehne CF, Lazo PA, Alves K, Lee JS, Tsichlis PN, O'Donnell PV (1989) The *Mlvi*-1 locus involved in the induction of rat T cell lymphomas and the *pvt*-1/*mis*-1 locus are identical. J Virol 63: 2366–2369

Kohl NE, Kanda N, Schreck RR, Bruns G, Latt SA, Gilbert F, Alt FW (1983) Transposition and amplification of oncogene-related sequences in human neuroblastomas. Cell 35: 359–367

Koizumi S, Fisher RJ, Fujiwara S, Jorcyk C, Bhat NK, Seth A, Papas TS (1990) Isoforms of the human *ets*-1 protein: generation by alternative splicing and differential phosphorylation. Oncogene 5: 675–681

Kozak CA (1985) Analysis of wild-derived mice for *Fv*-1 and *Fv*-2 murine leukemia virus restriction loci: a novel wild mouse *Fv*-1 allele responsible for lack of host range restriction. J Virol 55: 281–285

Kozak CA, Sears JF, Hoggan MD (1983) Genetic mapping of the mouse oncogenes c- Ha-*ras*-1 and c-*fes* to chromosome 7. J Virol 47: 217–220

Kozak CA, Strauss PG, Tsichlis PN (1985) Genetic mapping of a cellular DNA region involved in the induction of thymic lymphomas (*Mlvi*-1) to mouse chromosome 15. Mol Cell Biol 5: 894–897

Kumar V, Resnick P, Eastcott JW, Bennett M (1978a) Mechanism of genetic resistance to Friend virus leukemia in mice. V. Relevance of *Fv*-3 gene in the regulation in in vivo immunosuppression. JNCI 61: 1117–1123

Kumar V, Goldschmidt L, Eastcott JW, Bennett M (1978b) Mechanisms of genetic resistance to Friend virus leukemia in mice. IV. Identification of a gene (*Fv*-3) regulating immunosuppression in vitro, and its distinction from *Fv*-2 and genes regulating marrow allograft reactivity. J Exp Med 147: 422–433

Lacal JC, Moscat J, Aaronson SA (1987) Novel source of 1,2-diacylglycerol elevated in cells transformed by Ha-*ras* oncogene. Nature 330: 269

Laimins LA, Gruss P, Pozatti R, Khoury G (1984a) Characterization of the enhancer elements in the long terminal repeat of Moloney murine leukemia virus. J Virol 49: 183–189

Laimins L, Tsichlis PN, Khoury G (1984b) Multiple enhancer domains in the 3' terminus of the Prague strain of Rous sarcoma virus. Nucl Acid Res 12: 6427–6442

Lane DP, Crawford LV (1979) T antigen is bound to host protein in SV40-transformed cells. Nature 278: 261–263

Lane DP, Benchimol S (1990) p53: oncogene or anti-oncogene. Gene Dev 4: 1–8

Langdon WY, Hartley JW, Klinken SP, Ruscetti SK, Morse HC III (1989) V-*cbl*, an oncogene from a dual-recombinant murine retrovirus that induces early B cell lymphomas. Proc Natl Acad Sci USA 86: 1168–1172

Lazo PA, Tsichlis PN (1988) Recombination between two integrated proviruses one of which was inserted near c-*myc* in a retrovirus induced rat thymoma: implications for tumor progression. J Virol 62: 788–794

Lazo PA, Tsichlis PN (1990) Biology and pathogenesis of retroviruses. Semin Oncol 17: 269–294

Lazo PA, Klein-Szanto AJP, Tsichlis PN (1990a) T cell lymphoma lines derived from rat thymomas induced by Moloney murine leukemia virus: phenotypic diversity and its implications. J Virol 64: 3948–3959

Lazo PA, Lee JS, Tsichlis PN (1990b) Long distance activation of the c-*myc* proto-oncogene by provirus insertion in *Mlvi*-1 or *Mlvi*-4 in rat T-cell lymphomas. Proc Natl Acad Sci USA 87: 170–173

Leary AG, Yang YC, Clark SC, Gasson JC, Golde DW, Ogawa M (1987) Recombinant gibbon interleukin-3 supports formation of human multilineage colonies and blast cell colonies in culture: comparison with recombinant human granulocyte-macrophage colony-stimulating factor. Blood 70: 1343–1348

Le Beau MM, Westbrook CA, Diaz MO, Larson RA, Rowley JD, Gasson JC, Golde DW, Sherr CJ (1986) Evidence for the involvement of *GM-CSF* and *FMS* in the deletion (5q) in myeloid disorders. Science 231: 984–987

Ledbetter JA, June CH, Rabinovitch PS, Grossmann A, Tsu TT, Imboden JB (1988) Signal transduction through CD4 receptors: stimulatory vs. inhibitory activity is regulated by CD4 proximity to the CD3/T cell receptor. Eur J Immunol 18: 525–532

Lee JC, Ihle JN (1981a) Increased response to lymphokines are correlated with preleukemia in mice inoculated with Moloney leukemia virus. Proc Natl Acad Sci USA 78: 7712–7716

Lee JC, Ihle JN (1981b) Chronic immune stimulation is required for Moloney leukaemia virus-induced lymphomas. Nature 289: 407–409

Lee J, Mehta K, Blick MB, Gutterman JU, Lopez-Berestein G (1987) Expression of c-*fos*, c-*myb*, and c-*myc* in human monocytes: correlation with monocytic differentiation. Blood 69: 1542–1545

LeGouy E, DePinho R, Zimmerman D, Ferrier P, Collum R, Alt FW (1987) Structure and expression of *myc*-family genes. In: Alt FW, Harlow E, Ziff EB (eds) Nuclear Oncogenes. Cold Spring Harbor Laboratory Press, New York, pp 144–151

Lenz J, Celander D, Crowther RL, Patarca R, Perkins DW, Haseltine WA (1984) Determination of the leukaemogenicity of a murine retrovirus by sequences within the long terminal repeat. Nature (London) 308: 467–470

Leprince D, Duterque-Coquilland M, Ruo-Ping Li, Henry C, Flaurens A, Debuire B, Stehelin D (1988) Alternative splicing within the chicken c-*ets*-1 locus: Implications for transduction within the E26 retrovirus of the c-*ets* protooncogene. J Virol 62: 3233–3241

Leung K, Nabel GJ (1988) HTLV-I *trans*-activator induces interleukin-2 receptor expression through an NF-KB-like factor. Nature 333: 776–778

Levy DE, Lerner RA, Wilson MC (1985) The *Gv*-1 locus coordinately regulates the expression of multiple endogenous murine retroviruses. Cell 41: 289–299

Li JP, D'Andrea AD, Lodish HF, Baltimore D (1990) Activation of cell growth by binding of Friend spleen-focus-forming virus gp55 glycoprotein to the erythropoietin receptor. Nature 343: 762–764

Li Y, Holland CA, Hartley JW, Hopkins N (1984) Viral integration near c-*myc* in 10–20% of MCF247-induced AKR lymphomas. Proc Natl Acad Sci USA 81: 6808–6811

Li Y, Golemis E, Hartley JW, Hopkins N (1987) Disease specificity of non-defective Friend and Moloney murine leukemia viruses is controlled by a small number of nucleotides. J Virol 61: 693–700

Lilly F (1970) *Fv*-2: Identification and location of a second gene governing the spleen focus response to Friend leukemia virus in mice. J Natl Cancer Inst 45: 163–169

Lilly F, Pincus T (1973) Genetic control of murine viral leukemogenesis. Adv Cancer Res 64: 231–277

Lilly F, Boyse EA, Old LJ (1964) Genetic basis of susceptibility to viral leukemogenesis. Lancet 2: 1207–1209

Linzer DIH, Levine AJ (1979) Characterization of a 54K dalton cellular SV40 tumor antigen present in SV40 transformed cells and uninfected embryonal carcinoma cells. Cell 17: 43–52

Lonai P, Katz E, Peled A, Haran-Ghera N (1981) H-21-linked control of immunological resistance to viral leukemogenesis as a response to preleukemic cells. Immunogenetics 12: 423–432

Lopez AF, To LB, Yang YC, Gamble JR, Shannon MF, Burns GF, Dyson PG, Juttner CA, Clark S, Vadas MA (1987) Stimulation of proliferation, differentiation, and function of human cells by primate interleukin 3. Proc Natl Acad Sci USA 84: 2761–2765

LoSardo JE, Cupelli LA, Short MK, Berman JW, Lenz J (1989) Differences in activities of murine retroviral long terminal repeats in cytotoxic T lymphocytes and T lymphoma cells. J Virol 63: 1087–1094

LoSardo JE, Boral AL, Lenz J (1990) Relative importance of elements within the SL3-3 virus enhancer for T cell specificity. J Virol (in press)

Lowenthal JW, Bohnlein E, Ballard DW, Greene WC (1988) Regulation of interleukin 2 receptor α subunit (Tac or CD25 antigen) gene expression: binding of inducible nuclear proteins to discrete promoter sequences correlates with transcriptional activation. Proc Natl Acad Sci USA 85: 4468–4472

Lowenthal JW, Ballard DW, Bogerd H, Böhnlein E, Greene, WC (1989a) Tumor necrosis factor-β activation of the IL-2 receptor-α gene involves the induction of κB-specific DNA binding proteins. J Immunol 142: 3121–3128

Lowenthal JW, Ballard DW, Bohnlein E, Green WC (1989b) Tumor necrosis factor α induces proteins that bind specifically to κB-like enhancer elements and regulate interleukin-2 receptor β-chain gene expression in primary human T lymphocytes. Proc Natl Acad Sci USA 86: 2331–2335

Lung ML, Hartley JW, Rowe WP, Hopkins NH (1983) Large RNase T₁- resistant oligonucleotides encoding p15E and the U3 region of the long terminal repeat distinguish two biological classes of mink cell focus-forming type C viruses of inbred mice. J Virol 45: 275–290

Lusso PE, Ensoli B, Markham PD, Ablashi DV, Salahuddin SZ, Tschachler E, Wong- Staal F, Gallo RC (1989) Productive dual infection of human CD4⁺ T-lymphocytes by HIV-1 and HHV-6. Nature 337: 370–373

Lusso P, DiMarzo-Veronese F, Ensoli B, Franchini G, Jemma C, DeRocco SE, Kalyanaraman VS, Gallo RC (1990) Expanded HIV-1 cellular tropism by phenotypic mixing with murine endogenous retroviruses. Science 247: 848–852

Macara IG (1989) Elevated phosphorylcholine concentration in *ras*- transformed NIH 3T3 cells arises from increased choline kinase activity, not from phosphatidylcholine breakdown. Mol Cell Biol 9: 325–328

Magee AI, Gutierrez L, McKay IA, Marshall CJ, Hall A (1987) Dynamic fatty acylation of p21 N-*ras*. EMBO J 6: 3353–3357

Manager R, Najita L, Nichols EJ, Hakomori S, Rohrschneider L (1984) Cell surface expression of the McDonough strain of feline sarcoma virus *fms* gene product (gp^{fms}). Cell 39: 327–337

Manley NR, O'Connell MA, Sharp PA, Hopkins N (1989) Nuclear factors that bind to the enhancer region of nondefective Friend murine leukemia virus. J Virol 63: 4210–4223

Mann R, Mulligan RC, Baltimore D (1983) Construction of a retrovirus packaging mutant and its use to produce helper-free defective retrovirus. Cell 33: 153–159

Manne V, Bekesi E, Kung HF (1985) Ha-*ras* proteins exhibit GTPase activity: point mutations that activate Ha-*ras* products result in decreased GTPase activity. Proc Natl Acad Sci USA 82: 376–380

Marshall CJ, Lloyd AC, Morris JDH, Paterson H, Price B, Hall A (1989) Signal transduction by p21*ras*. Int J Cancer (Supplement) 4: 29–31

Marth JD, Peet R, Krebs EG, Perlmutter RM (1985) A lymphocyte-specific protein-tyrosine kinase gene is rearranged and overexpressed in the murine T cell lymphoma LSTRA. Cell 43: 393–404

Marth JD, Disteche C, Pravtcheva D, Ruddle F, Krebs EG, Perlmutter RM (1986) Localization of a lymphocyte-specific protein tyrosine kinase gene (LCK) at a site of frequent chromosomal abnormalities in human lymphomas. Proc Natl Acad Sci USA 83: 7400–7404

Maruyama M, Shibuya H, Harada M, Hatakeyama M, Seiki M, Fujita T, Inoue J, Yoshida M, Taniguchi T (1987) Evidence for aberrant activation of the interleukin-2 loop by HTLV-I encoded p40^X and T3/Ti complex triggering. Cell 48: 343–350

McCarter JA, Ball JK, Frei JV (1977) Lower limb paralysis induced in mice by a temperature-sensitive mutant of Moloney leukemia virus. JNCI 59: 179–183

McCormick F (1989) *ras* GTPase activity protein: signal transmitter and signal terminator. Cell 56: 5–8

Meier H, Myers DD, Huebner RJ (1969) Genetic control by the *hr*- locus of susceptibility and resistance to leukemia. Proc Natl Acad Sci USA 63: 759–766

Mercer WE, Nelson D, Deleo AB, Old LJ, Baserga R (1982) Microinjection of monoclonal antibody to protein p53 inhibits serum-induced DNA synthesis in 3T3 cells. Proc Natl Acad Sci USA 79: 6309–6312

Mercer WE, Avignolo C, Baserga R (1984) Role of the p53 protein in cell proliferation as studied by microinjection of monoclonal antibodies. Mol Cell Biol 4: 276–281

Meruelo D (1979) A role for elevated H-2 antigen expression in resistance to neoplasia caused by radiation-induced leukemia virus. Enhancement of effective tumor surveillance by killer lymphocytes. J Exp Med 149: 898–909

Meruelo D, Bach R (1983) Genetics of resistance to virus-induced leukemias. Adv Cancer Res 40: 107–188

Meruelo D, Lieberman M, Ginzton N, Deak B, McDevitt HO (1977) Genetic control of radiation leukemia virus-induced tumorigenesis. J Exp Med 146: 1079–1087

Meyerhans A, Cheynier R, Albert J et al. (1989) Temporal fluctuations in HIV quasi species in vivo are not reflected by sequential HIV isolations. Cell 58: 901–910

Miles BD, Robinsons HL (1985) High-frequency transduction of c-*erb*B in avian leukosis virus-induced erythroblastosis. J Virol 54: 295–303

Moelling K, Pfaff E, Beug H, Beimling P, Bunte T, Schaller HE, Graf T (1985) DNA-binding activity is associated with purified Myb proteins from AMV and E26 viruses and is temperature-sensitive for E26 ts mutants. Cell 40: 983–990

Montminy MR, Sevarino KA, Wagner JA, Mandel G, Goodman RH (1986) Identification of a cyclic-AMP-responsive element within the rat somatostatin gene. Proc Natl Acad Sci USA 83: 6682–6686

Mooslehner K, Karls U, Harbers K (1990) Retroviral integration sites in transgenic Mov mice frequently map in the vicinity of transcribed DNA regions. J Virol 64: 3056–3058

Moreau-Gachelin F, Tavitian A, Tambourin P (1988) *Spi*-1 is a putative oncogene in virally induced murine erythroleukemias. Nature 331: 277–280

Moreau-Gachelin F, Ray D, Mattei GG, Tambourin P, Tavitian A (1989) The putative oncogene *Spi*-1: murine chromosomal localization and transcriptional activation in murine acute erythro-leukemias. Oncogene 4: 1449–1456

Morishita K, Parker DS, Mucenski ML, Jenkins NA, Copeland NG, Ihle JN (1988) Retroviral activation of a novel gene encoding a zinc finger protein in IL-3-dependent myeloid leukemia cell lines. Cell 54: 831–840

Morishita K, Parganas E, Bartholomew C, Sacchi N, Valentine MB, Raimondi SC, LeBeau MM, Ihle JN (1990) The human *Evi*-1 gene is located on chromosome 3q24-q28 but is not rearranged in three cases of acute nonlymphocytic leukemias containing t(3;5)(q25;q34) translocations. Oncogene Res 5: 221–231

Morrissey PJ, Parkinson DR, Schwartz RS, Waksal SD (1980) Immunologic abnormalities in HRS/J mice. I. Specific deficit in T-lymphocyte helper function in a mutant mouse. J Immunol 125: 1558–1562

Morse HC, Yetter RA, Via CS et al. (1989) Functional and phenotypic alterations in T cell subsets during the course of MAIDS, a murine retrovirus-induced immunodeficiency syndrome. J Immunol 143: 844–850

Mosca JD, Bednank NB, Raj NBK et al. (1987) Activation of human immunedeficiency virus by herpes virus infection: identification of a region within the long terminal repeat that responds to a transacting factor encoded by herpes simplex virus 1. Proc Natl Acad Sci USA 84: 7408–7412

Mowat M, Cheng A, Kimura N, Bernstein A, Benchimol S (1985) Rearrangements of the cellular p53 gene in erythroleukaemic cells transformed by Friend virus. Nature 314: 633–636

Mucenski ML, Taylor BA, Jenkins NA, Copeland NG (1986) AKXD recombinant inbred strains: models for studying the molecular genetic basis of murine lymphomas. Mol Cell Biol 6: 4236–4243

Mucenski ML, Gilbert DJ, Taylor BA, Jenkins NA, Copeland NG (1987a) Common sites of viral integration in lymphomas arising in AKXD recombinant inbred mouse strains. Oncogene Res 2: 33–48

Mucenski ML, Taylor BA, Copeland NG, Jenkins NA (1987b) Characterization of somatically acquired ecotropic and mink cell focus-forming viruses in lymphomas of AKXD recombinant inbred mice. J Virol 62: 2929–2933

Mucenski ML, Bedigian HG, Shull MM, Copeland NG, Jenkins NA (1988a) Comparative molecular genetic analysis of lymphomas from six inbred mouse strains. J Virol 62: 839–846

Mucenski ML, Taylor BA, Copeland NG, Jenkins NA (1988b) Chromosomal location of Evi-1, a common site of ecotropic viral integration in AKXD murine myeloid tumors. Oncogene Res 2: 219–233

Mucenski ML, Taylor BA, Ihle JN, Hartley JW, Morse III HC, Jenkins NA, Copeland NG (1988c) Identification of a common ecotropic viral integration site, evi-1, in the DNA of AKXD murine myeloid tumors. Mol Cell Biol 8: 301–308

Mullins JI, Brody DS, Binari RC Jr, Cotter SM (1983) Viral transduction of c-myc gene in naturally occurring feline leukemias. Nature 308: 856–858

Murray AW, Kirschner MW (1989) Cyclin synthesis drives the early embryonic cell cycle. Nature 339: 275–280

Murray AW, Solomon MJ, Kirschner MW (1989) The role of cyclin synthesis and degradation in the control of maturation promoting factor activity. Nature 339: 280–286

Murre C, McCaw PS, Baltimore D (1989a) A new DNA binding and dimerization motif in immunoglobulin enhancer binding, daughterless, MyoD, and myc proteins. Cell 56: 777–783

Murre C, McCaw PS, Vaessin H, Caudy M, Jan LY, Jan YN, Cabrera CV, Buskin JN, Hauschka SD, Lassar AB, Weintraub H, Baltimore D (1989b) Interactions between heterologous helix-loop-helix proteins generate complexes that bind specifically to a common DNA sequence. Cell 58: 537–544

Mushinsky JF, Potter M, Bauer SR, Reddy EP (1983) DNA rearrangement and altered RNA expression of the c-myb oncogene in mouse plasmocytoid lymphosarcoma. Science 220: 795–798

Nagarajan L, Louie E, Tsujimoto Y, ar-Rushdi A, Huebner K, Croce CM (1986) Localization of the human pim oncogene (PIM) to a region of chromosome 6 involved in translocations in acute leukemias. Proc Natl Acad Sci USA 83: 2556–2560

Nagata K, Ohtani K, Nakamura M, Sugamura K (1989) Activation of endogenous c-fos proto-oncogene expression by human T cell leukemia virus type I encoded p40^{tax} protein in the human T-cell line, Jurkat. J Virol 63: 3220–3226

Neel BG, Hayward WS, Robinson HL, Fan J, Astrin SM (1981) Avian leukosis virus-induced tumors have common proviral integration sites and synthesize discrete new RNAs: oncogenesis by promoter insertion. Cell 23: 323–334

Neil JC, Hughes R, McFarlane R, Wilkie MM, Onions DE, Lees G, Jarrett O (1983) Transduction and rearrangement of the myc gene by feline leukemia virus in naturally occurring T cell leukemias. Nature 308: 814–820

Nerenberg M, Hinrichs S, Reynolds RK, Khoury G, Jay G (1987) The tat gene of human T lymphotropic virus type 1 induces mesenchymal tumors in transgenic mice. Science 237: 1324–1329

Nichols EJ, Manger R, Hakomori S, Rohrschneider LR (1987) Transformation by the oncogene v-fms: the effects of castanospermine on transformation-related parameters. Exp Cell Res 173: 486–495

Nimer SD, Gasson JC, Hu K, Smalberg I, Williams JL, Chen ISY, Rosenblatt JD (1989) Activation of the GM-CSF promoter by HTLV-I and II tax proteins. Oncogene 4: 671–676

Nottenburg C, Stubblefield E, Varmus HE (1987) An aberrant avian leukosis virus provirus inserted downstream from the chicken c-myc coding sequence in a bursal lymphoma results from intrachromosomal recombination between two proviruses and deletion of cellular DNA. J Virol 61: 1828–1833

Nunn M, Weiher H, Bullock P, Duesberg P (1984) Avian erythroblastosis virus E26: nucleotide sequence of the tripartite onc gene and of the LTR and analysis of the cellular prototype of the viral ets sequence. Virology 139: 330–339

Odaka T (1970) Inheritance of susceptibility to Friend mouse leukemia virus. VII. Establishment of a resistant strain. Int J Cancer 6: 18–23

Odaka T, Ikeda H, Moriwaki K, Matsuzawa A, Mizuno M, Kondo K (1978) Genetic resistance in Japanese wild mice (Mus musculus molossinus) to an NB-tropic Friend murine leukemia virus. J Natl Cancer Inst 61: 1301–1306

Odaka T, Ikeda H, Yoshikura H, Moriwaki K, Suzuki S (1981) Fv-4 : gene controlling resistance to NB-tropic Friend murine leukemia virus. Distribution in wild mice, introduction into genetic background of BALB/c mice, and mapping of chromosomes. J Natl Cancer Inst 67: 1123–1127

O'Donnell PV, Woller R, Chu A (1984) Stages in development of mink cell focus- inducing (MCF) virus-accelerated leukemia in AKR mice. J Exp Med 160: 914–934

O'Donnell PV, Fleissner E, Lonial H, Koehne CF, Reicin A (1985) Early clonality and high-frequency proviral integration into the c-myc locus in AKR leukemias. J Virol 55: 500–503

Oliff A, Signorelli K, Collins L (1984) The envelope gene and long terminal repeat sequences contribute to the pathogenic phenotype of helper-independent Friend viruses. J Virol 51: 788–794

Oliff A, McKinney MD, Agranovsky O (1985) Contribution of the gag and pol sequences to the leukemogenicity of Friend murine leukemia virus. J Virol 54: 864–868

O'Neill HC, McGrath MS, Allison JP, Weissman IL (1987) A subset of T-cell receptors associated with L3T4 molecules mediates C6VL leukemia cell binding of its cognate retrovirus. Cell 49: 143–151

O'Neill RR, Khan AS, Hoggan MD, Hartley JW, Martin MA, Repaske R (1986) Specific hybridization probes demonstrate fewer xenotropic than mink cell focus-forming murine leukemia virus env-related sequences in DNAs from inbred laboratory mice. J Virol 53: 359–366

Papkoff J, Nigg EA, Hunter T (1983) The transforming protein of Moloney murine sarcoma virus is a soluble cytoplasmic protein. Cell 33: 161–172

Parada LF, Land H, Weinberg RD, Wolf D, Rotter V (1984) Cooperation between gene encoding p53 tumor antigen and ras in cellular transformation. Nature 312: 649–651

Paskalis H, Felber BK, Pavlakis GN (1986) Cis acting sequences responsible for the transcriptional activation of human T cell leukemia virus type I constitute a conditional enhancer. Proc Natl Acad Sci USA 83: 6558–6562

Paul R, Schuetze S, Kozak L, Kabat D (1989) A common site for immortalizing proviral integrations in Friend erythroleukemia: molecular cloning and characterization. J Virol 63: 4958–4961

Paul R, Schuetze S, Kozak SL, Kozak CA, Kabat DA (1990) The Sfpi-1 proviral integration site of Friend erythroleukemia encodes the Ets-1 related transcription actor PU.1 and a small RNA. J Virol (in press)

Peters G, Lee AE, Dickson C (1986) Concerted activation of two potential proto-oncogenes in carcinomas induced by mouse mammary tumor virus. Nature 320: 628–631

Peters G, Brookes S, Smith R, Placzek M, Dickson C (1989) The mouse homolog of the hst/k-FGF gene is adjacent to int-2 and is activated by proviral insertion in some virally induced mammary tumors. Proc Natl Acad Sci USA 86: 5678–5682

Pettenati MJ, Le Beau MM, Lemons RS, Shima EA, Kawasaki ES, Larson RA, Sherr C, Daz MO, Rowley JD (1987) Assignment of CSF-1 to 5q33.1: evidence for clustering of genes regulating hematoposoiesis and for their involvement in the deletion of the long arm of chromosome 5 in myeloid disorders. Proc Natl Acad Sci. USA 84: 2970–2974

Phillips NE, Parker DC (1987) Fc-γ receptor effects on induction of c-myc mRNA expression in mouse B lymphocytes by anti-immunoglobulin. Mol Immunol 24: 1199–1205

Pincus T, Hartley JW, Rowe WP (1971a) A major genetic locus affecting resistance to infection with murine leukemia viruses. I. Tissue culture studies of naturally occurring viruses. J Exp Med 133: 1219–1233

Pincus T, Rowe WP, Lilly F (1971b) A major genetic locus affecting resistance to infection with murine leukemia viruses. II. Apparent identity to a major locus described for resistance to Friend murine leukemia virus. J Exp Med 133:1234–1241

Plata F (1982) Specificity studies of cytolytic T-lymphocytes directed against murine leukemia virus-induced tumors. Analysis by monoclonal cytolytic T lymphocytes. J Exp Med 155: 1050–1062

Plata F, Lilly F (1979) Viral specificity of H-2-restricted T-killer cells directed against syngeneic tumors induced by Gross, Friend, or Rauscher leukemia virus. J Exp Med 150: 1174–1186

Poe M, Scolnick EM, Stein RB (1985) Harvey ras p21 expressed in Escherichia coli purifies as a binary one-to-one complex with GDP. J Biol Chem 260: 3906–3909

Pognonec P, Boulukos KE, Gesquiere JC, Stehelin D, Ghysdael J (1988) Mitogenic stimulation of thymocytes results in the calcium-dependent phosphorylation of c-ets-1 proteins. EMBO J 7: 977–983

Pognonec P, Boulukos KE, Ghysdael J (1989) The c-ets-1 protein is chromatin associated and binds to DNA in vitro. Oncogene 4: 691–697

Pognonec P, Boulukos KE, Bosselut R, Boyer C, Schmitt-Verhulst AM, Ghysdael J (1990) Identification of a *Ets1* variant protein unaffected in its chromatin and in vitro DNA binding capacities by T cell antigen receptor triggering and intracellular calcium rises. Oncogene 5: 603–610

Poirier Y, Kozak C, Jolicoeur P (1988) Identification of a common helper provirus integration site in Abelson murine leukemia virus-induced lymphoma DNA. J Virol 62: 3985–3992

Popovic M, Lange-Wantzin G, Sarin PS, Mann D, Gallo RC (1983) Transformation of human umbilical cord blood T cells by human T cell leukemia/lymphoma virus. Proc Natl Acad Sci USA 80: 5402–5406

Poteat HT, Kadison P, McGuire K, Park L, Park RE, Sodroski JG, Haseltine WA (1989) Response of the human T cell leukemia virus type 1 long terminal repeat to cyclic AMP. J Virol 63: 1604–1611

Poteat HT, Chen FY, Kadison P, Sodroski JG, Haseltine WA (1990) Protein kinase A-dependent binding of a nuclear factor to the 21-base-pair repeat of the human T cell leukemia virus type I long terminal repeat. J Virol 64: 1264–1270

Propst F, Rosenberg MP, Iyer A, Kaul K, Vande Woude GF (1987) c-*mos* proto-oncogene RNA transcripts in mouse tissues: structural features, developmental regulation, and localization in specific cell types. Mol Cell Biol 7: 1629–1637

Propst F, Vande Woude GF, Jenkins NA, Copeland NG, Lee BK, Hunt PA, Eicher EM (1989) The *mos* proto-oncogene maps near the centromere on mouse chromosome 4. Genomics 5: 118–123

Quint W, Boelens W, van Wezenbeek P, Cuypers T, Maandag ER, Selten G, Berns A (1984) Generation of AKR mink cell focus-forming viruses: a conserved single copy xenotropic-like provirus provides recombinant long terminal repeat sequences. J Virol 50: 432–438

Rabbitts TH, Foster A, Hamlyn P, Baer R (1984) Effect of somatic mutation within translocated c-*myc* genes in Burkitt's lymphoma. Nature (London) 309: 592–597

Rapp UR, Goldsborough MD, Mark GE, Bonner TI, Groffen J, Reynolds FH Jr, Stephenson JR (1983) Structure and biological activity of v-*raf*, a unique oncogene transduced by a retrovirus. Proc Natl Acad Sci USA 80: 4218–4222

Ratner L, Poiesz BJ (1988) Leukemias associated with human T cell lymphotropic virus type I in non-endemic region. Medicine 67: 401–422

Rauscher FJ III, Cohen DR, Curran T, Bos TJ, Vogt PK, Bohmann D, Tjian R, Franza RB Jr (1988) *Fos*-associated protein p39 is the product of the *jun* proto-oncogene. Science 240: 1010–1016

Rayter SI, Iwata KK, Michitsch RW, Sorvill JM, Valenzuela DM, Foulkes JG (1989) Biochemical functions of oncogenes. In: Clover DM, Hanes BD (eds) Oncogenes. IRL Press, Oxford, pp113–189

Rechavi G, Givol D, Canaani E (1982) Activation of a cellular oncogene by DNA rearrangement: possible involvement of an IS-like element. Nature 300: 607–611

Reed JC, Alpers JD, Nowell PC, Hoover RG (1986) Sequential expression of protooncogenes during lectin-stimulated mitogenesis of normal human lymphogytes. Proc Natl Acad Sci USA 83: 3982–3986

Reddy ES, Rao VN (1988) Structure, expression and alternative splicing of the human c-*ets*-1 proto-oncogene. Oncogene Res 3: 239–246

Reicin A, Yang JQ, Marcu KB, Fleissner E, Koehne CF, O'Donnell PV (1986) Deregulation of the c-*myc* oncogene in virus-induced thymic lymphomas of AKR/J mice. Mol Cell Biol 6: 4088–4092

Rice AP, Mathews MB (1988) Trans-activation of the human immunodeficiency virus long terminal repeat sequences, expressed in an adenovirus vector by the adenovirus E1A 13S protein. Proc Natl Acad Sci USA 85: 4200–4204

Rimsky L, Hauber J, Dukovich M et al. (1988) Functional replacement of HIV-1 *rev* protein by the HTLV-1 *rex* protein. Nature 335: 738–740

Risser R (1982) The pathogenesis of Abelson virus lymphomas in the mouse. Biochim Biophys Acta 651: 213–244

Roberts L (1990) Down to the wire for the NF gene. Science 249: 236–238

Robinson HL, Gagnon GC (1986) Patterns of proviral insertion and deletion in avian leukosis virus-induced lymphomas. J Virol 57:28–36

Robinson HL, Pearson MN, De Simone DW, Tsichlis PN, Coffin JM (1979) Subgroup E avian leukosis virus associated disease in chickens. Cold Spring Harbor Symp Quant. Biol 44: 1133–1142

Robinson HL, Blais BM, Tsichlis PN, Coffin JM (1982) Two regions of the viral genome determine the oncogenic potential of avian leukosis viruses. Proc Natl Acad Sci USA 79: 1225–1229

Rohdewohld H, Weiher H, Reik W, Jaenisch R (1987) Retrovirus integration and chromatin structure: Moloney murine leukemia proviral integration sites map near DNase I-hypersensitive sites. J Virol 61: 336–343

Rosen CA, Haseltine WA, Lenz J, Ruprecht R, Cloyd MW (1985) Tissue selectively of murine leukemia virus infection is determined by long terminal repeat sequences. J Virol 55: 862–866

Rosson D, Dugan D, Reddy EP (1987) Aberrant splicing events that are induced by proviral integration: implication for myb oncogene activation. Proc Natl Acad Sci USA 84: 3171–3175

Rothwell VM, Rohrschneider LR (1987) Murine c-fms cDNA: cloning, sequence analysis and retroviral expression. Oncogene Res 1: 311–324

Rotter V, Wolf D, Pravtcheva D, Ruddle F (1984) Chromosomal assignment of the murine gene encoding the transformation related protein p53. Mol Cell Biol 4: 383–385

Roussel MF, Rettenmier CW, Look AT, Sherr CJ (1984) Cell surface expression of v-fms coded glycoproteins is required for transformation. Mol Cell Biol 4: 1999–2009

Roussel MF, Downing JR, Rettemier CW, Sherr CJ (1988) A point mutation in the extracellular domain of the human CSF-1 receptor (c-fms protooncogene product) activates its transforming potential. Cell 55: 979–988

Rovinski B, Munroe D, Peacok M, Mowat M, Bernstein A, Benchimol S (1987) Deletion of 5' coding sequences of the cellular p53 gene in mouse erythroleukemia: a novel mechanism of oncogene regulation. Mol Cell Biol 7: 847–853

Rowe WP, Humphrey JB, Lilly F (1973) A major genetic locus affecting resistance to infection with murine leukemia viruses. III. Assignment of the Fv-1 locus to linkage group VIII of the mouse. J Exp Med 173: 850–853

Rowe WP, Cloyd MW, Hartley JW (1979) Status of the association of mink cell focus-forming viruses with leukemogenesis. Cold Spring Harbor Symp Quant Biol 44: 1265–1268

Ruben S, Poteat H, Tan TH et al. (1988) Cellular transcription factors and regulation of IL-2 receptor gene expression by HTLV-1 tax gene. Science 241: 89–92

Ruegg CL, Monell CR, Strand M (1989) Identification, using synthetic peptides, of the minimum amino acid sequence from the retroviral transmembrane protein p15E required for inhibition of lymphoproliferation and its similarity to gp 21 of human T lymphotropic virus types I and II. J Virol 63: 3250–3256

Ruscetti S, Davis L, Feild J, Oliff A (1981) Friend murine leukemia virus-induced leukemia is associated with the formation of mink cell focus-inducing viruses · and is blocked in mice expressing endogenous mink cell focus-inducing xenotropic viral envelope genes. J Exp Med 154: 907–920

Sacchi N, Watson DK, Geurts van Kessel AHM, Hagemeijer A, Kersey J, Drabkin HD, Patterson D, Papas TS (1986) Hu-ets-1 and Hu-ets-2 genes are transposed in acute leukemias with (4; 11) and (8; 21) translocations. Science 231: 379–381

Sagata N, Oskarsson M, Copeland T, Brumbauch J, Vande Woude GF (1988) Function of c-mos proto-oncogene product in meiotic maturation in Xenopus oocytes. Nature 335: 519–525

Sagata N, Watanabe N, Vande Woude GF, Ikawa Y (1989) The c-mos proto-oncogene product is a cytostatic factor responsible for meiotic arrest in vertebrate eggs. Nature 342: 512–518

Saizawa K, Rojo J, Janeway CA (1987) Evidence for a physical association of CD4 and the CD3: $\alpha : \beta$ T cell receptor. Nature 328: 260–263

Sakaguchi AY, Lalley PA, Zabel BU, Ellis RW, Scolnick EM, Naylor SL (1984) Chromosome assignments of four mouse cellular homologs of sarcoma and leukemia virus oncogenes. Proc Natl Acad Sci USA 81: 525–529

Sarnow P, Ho YS, Williams J, Levine AJ (1982) Adenovirus E1b-58kd tumor antigen and SV40 large tumor antigen are physically associated with the same 54kd cellular protein in transformed cells. Cell 28: 387–394

Sassone-Corsi P, Ransone LJ, Lamph WW, Verma IM (1988) Direct interaction between fos and jun nuclear oncoproteins: role of the leucine zipper domain. Nature 336: 692–695

Schüpbach J (1989) Human retrovirology: facts and concepts. Cur Top Microbiol Immunol 142: 1–115

Scolnick EM, Papageorge AG, Shih TY (1979) Guanine nucleotide-binding activity as an assay for src protein of rat-derived murine sarcoma viruses. Proc Natl Acad Sci USA 76: 5355–5359

Seiki M, Eddy R, Show TB et al. (1984) Non specific integration of the HTLV provirus genome into adult T cell leukemia cells. Nature 309: 640–642

Selten G, Cuypers HT, Zijlstra M, Melief C, Berns A (1984) Involvement of c-myc in MuLV-induced T cell lymphomas in mice: frequency and mechanism of activation. EMBO J 3: 3215–3222

Selten G, Cuypers HT, Berns A (1985) Proviral activation of the putative oncogene Pim-1 in MuLV induced T cell lymphomas. EMBO J 4: 1793–1798

Selten G, Cuypers HT, Boelens W, Robanus-Maandag E, Verbeek J, Domen J, Van Beveren C, Berns A (1986) The primary structure of the putative oncogene pim-1 shows extensive homology with protein kinases. Cell 46: 603–611

Sen, R, Baltimore D (1986) Inducibility of κ in Immunoglobulin enhancer-binding protein NF-κB by a posttranslational mechanism. Cell 47: 921–928

Seto E, Yen TS, Peterlin BM et al (1988) Transactivation of the human immunodeficiency virus long terminal repeat by the hepatitis B virus X protein. Proc Natl Acad Sci USA 85: 8286–8290

Setoguchi M, Higuchi Y, Yoshida S, Nasu N, Miyazaki Y, Akizuki S, Yamamoto S (1989) Insertional activation of N-myc by endogenous Moloney-like murine retrovirus sequences in macrophage cell lines derived from myeloma cell line-marcrophage hybrids. Mol Cell Biol 9: 4515–4522

Shaw G, Kamen R (1986) A conserved AU sequence from the 3′ untranslated region of GM-CSF mRNA mediates selective mRNA degradation. Cell 46: 659–667

Sheiness D, Gardinier M (1984) Expression of a proto-oncogene (proto-myb) in hemopoietic tissues of mice. Mol Cell Biol 4: 1206–1212

Shen-Ong GLC, Potter M, Mushinsky JF, Lavu S, Reddy EP (1984) Activation of the c-myb locus by viral insertional mutagenesis in plasmocytoid lymphosarcomas. Science 226: 1077–1080

Shen-Ong GLC, Morse HC III, Potter M, Mushinski F (1986) Two modes of c-myb activation in virus-induced mouse myeloid tumors. Mol Cell Biol 6: 380–392

Sherr CJ (1988) The fms oncogene. Biochim Biophys Acta 948: 225–243

Sherr CJ, Rettenmier CW, Sacca R, Roussel MF, Look AT, Stanley ER(1985) The c-fms proto-oncogene product is related to the receptor for the mononuclear phagocyte growth factor, CSF-1. Cell 41: 665–676

Shibuya T, Mak TW (1982a) A host gene controlling early anemia or polycythemia induced by Friend erythroleukemia virus. Nature 296: 577–579

Shih CC, Stoye JP, Coffin JM (1988) Highly preferred targets for retrovirus integration. Cell 53: 531–537

Shobat O, Greenberg M, Reisman D, Oren M, Rotter V (1987) Inhibition of cell growth mediated by plasmids encoding p53 antisense. Oncogene 1: 277–283

Short MK, Okenquist SA, Lenz J (1987) Correlation of leukemogenic potential of murine retroviruses with transcriptional preference of the viral long terminal repeats. J Virol 61: 1067–1072

Sieff CA, Emerson SG, Donahue RE, Nathan DG (1985) Human recombinant granulocyte-macrophage colony-stimulating factor. A multilineage hematopoietin. Science 230: 1171–1173

Sieff CA, Niemeyer CM, Nathan DG, Ekern SC, Bieber FR, Yang YC, Wong G, Clark SC (1987) Stimulation of human hematopoietic colony formation by recombinant gibbon multi-colony-stimulating factor or interleukin 3. J Clin Invest 80: 818–823

Siekevitz M, Feinberg MB, Holbrook N, Wong-Staal F, Greene WC (1987) Activation of interleukin-2 and interleukin-2 receptor (Tac) promoter expression by the trans activator (tat) gene product of human T cell leukemia virus, type I. Proc Natl Acad Sci USA 84: 5389–5393

Silver BJ, Bokar JA, Virgin JB, Vallen EA, Milsted A, Wilson JH (1987) Cyclic AMP regulation of the human glycoprotein hormone α subunit gene is mediated by an 18 base pair element. Proc Natl Acad Sci USA 84: 2198–2202

Silver J, Buckler CE (1986) A preferred region for integration of Friend murine leukemia virus in hematopoietic neoplasms is closely linked to the int-2 oncogene. J Virol 60: 1156–1158

Silver J, Kozak C (1986) Common proviral integration region on mouse chromosome 7 in lymphomas and myelogenous leukemias induced by Friend murine leukemia virus. J Virol 57: 526 533

Silver J, Teich N (1981) Expression of resistance to Friend virus-stimulated erythropoiesis in bone marrow chimeras containing Fv-2^{rr} and Fv^{ss} bone marrow. J Exp Med 154: 126–137

Silvers WK (1979) White-spotting, patch and rump-white. In: The coat colors of mice: a model for gene action and interaction. Springer Berlin Heidelberg, New York, pp 206–241

Simpson RU, Hsu T, Begley DA, Mitchell BS, Alizadeh BN (1987) Transcriptional regulation of c-myc proto-oncogene by 1,25-dihydroxyvitamin D_3in HL-60 promyelocytic leukemia cells. J Biol Chem 262: 4104–4108

Singh B, Hannink M, Donoghue DJ, Arlinghaus RB (1986) p37^{mos}-associated serine/threonine protein kinase activity correlates with the cellular transformation function of v-mos. J Virol 60: 1148–1152

Singh B, Wittenberg C, Hannink M, Reed SI, Donoghue DJ, Arlinghaus RB (1988) The histidine-221 to tyrosine substitution in v-mos abolishes its biological function and its protien kinase activity. Virology 164: 114–120

Sola B, Fichelson S, Bordereaux D, Tambourin PE, Gisselbrecht S (1986) fim-1 and fim-2: two new integration regions of Friend murine leukemia virus in myeloblastic leukemias. J Virol 60: 718–725

Sola B, Simon S, Mattei MG, Fichelson S, Bordereaux D, Tambourin PE, Guenet JL, Gisselbrecht S (1988) *Fim*-1, *fim*-2/c-*fms*, and *fim*- 3, three common integration sites of Friend murine leukemia virus in myeloblastic leukemias, map to mouse chromosome 13, 18 and 3 respectively. J Virol 62: 3973–3978

Speck NA, Baltimore D (1987) Six different nuclear factors interact with the 75-base pair direct repeat of the Moloney murine leukemia virus enhancer. Mol Cell Biol 7: 1101–1110

Speck NA, Renjifo B, Hopkins N (1990a) Point mutations in the Moloney murine leukemia virus enhancer identify a lymphoid-specific viral core motif and 1,3-phorbol myristate acetate-inducible element. J Virol 64: 543–550

Speck NA, Renjifo B, Golemis E, Fredrickson TN, Hartley JW, Hopkins N (1990b) Mutation of the core or adjacent LVb elements of the Moloney murine leukemia virus enhancer alters disease specificity. Genes 4: 233–242

Spector DH, Wade E, Wright DA, Koval V, Clark C, Jaquish D, Spector SA (1990) Human immunodeficiency virus pseudotypes with expanded cellular and species tropism. J Virol 64: 2298–2308

Spector DL, Watt RA, Sullivan NF (1987) The v- and c-*myc* oncogene proteins colocalize in situ with small nuclear ribonucleoprotein particles. Oncogene 1: 5–12

Spiro C, Gliniak B, Kabat D (1988) A tagged helper-free Friend virus causes clonal erythroblast immortality by specific proviral integration in the cellular genome. J Virol 62: 4129–4135

Sporn MB, Roberts AB (1985) Autocrine growth factors and cancer. Nature 313: 745–747

Stacey DW, Kung HF (1984) Transformation of NIH 3T3 cells by microinjection of Ha-*ras* p21 protein. Nature 310: 508–511

Steffen D (1984) Proviruses are adjacent to c-*myc* in some murine leukemia virus-induced lymphomas. Proc Natl Acad Sci USA 81: 2097–2101

Steeves RA, Mirand EA, Bulba A, Trudel PJ (1970) Spleen foci and polycythemia in C57BL mice infected with host-adapted Friend leukemia virus. Int J Cancer 5: 346–356

Stockert E, Old LJ, Boyse EA (1971) The Gix System. A cell surface allo-antigen associated with murine leukemia virus; implications regarding chromosomal integration of the viral genome. J Exp Med 149: 200–215

Stockert E, Boyse EA, Obata Y, Ikeda H, Sarker NH, Hoffman HA (1975) New mutant and congenic mouse stocks expressing the murine leukemia virus- associated thymocyte surface antigen Gix. J Exp Med 142: 512–517

Stocking C, Löliger C, Kawai M, Suciu S, Gough N, Ostertag W (1988) Identification of genes involved in growth autonomy of hematopoietic cells by analysis of factor-independent mutants. Cell 53: 869–879

Storch TG, Chused TM (1984) Sex and H-2 haplotype control the resistance of CBA-BALB hybrids to the induction of T-cell lymphoma by Moloney leukemia virus. J Immunol 133: 2797–2800

Storch TG, Arnstein P, Monahar U, Leiserson WM, Chused TM (1985) Proliferation of infected lymphoid precursors before Moloney murine leukemia virus-induced T-cell lymphoma. JNCI 74: 137–143

Stoye J, Coffin J (1985) Endogenous retroviruses. In: Weiss R, Teich N, Varmus H et al. (eds) RNA tumor viruses, vol 2 (edn 2). Cold Spring Harbor, New York, pp 357–404

Stoye JP, Coffin JM (1987) The four classes of endogenous murine leukemia virus: structural relationships and potential for recombination. J Virol 61: 2659–2669

Stoye JP, Coffin JM (1988a) Polymorphism of murine endogenous proviruses revealed by using virus class-specific oligonucleotide probes. J Virol 62: 168–175

Stoye JP, Fenner S, Greenoak GE, Moran C, Coffin JM (1988b) Role of endogenous retroviruses as mutagens: the hairless mutation of mice. Cell 54: 383–391

Sturzbecher HW, Chumakov P, Welch WJ, Jenkins JR (1987) Mutant p53 proteins bind p53 hsp 72/73 cellular heat-shock-related proteins in SV40-transformed monkey cells. Oncogene 1: 201–211

Sumegi J, Spira J, Hazin H, Szpirer J, Levan G, Klein G (1983) Rat c-*myc* oncogene is located on chromosome 7 and rearranges in immunocytomas with t(6:7) chromosomal translocation. Nature 306: 497–498

Suzuki S (1975) *Fv-4*: a new gene affecting the splenomegaly induction by Friend leukemia virus. Jpn J Exp Med 45: 473–478

Suzuki S, Axelrad A (1980) *Fv-2* locus controls the proportion of erythropoietic progenitor cells (BFU-E) synthesizing DNA in normal mice. Cell 19: 225–236

Suzuki S, Tsuji K, Moriwaki K (1981) Friend murine leukemia virus resistance in Japanese wild mice: possible allelism with *Fv-4* in FRG mice. J Natl Cancer Inst 66: 729–731

Swain SL (1983) T cell subsets and the recognition of MHC class. Immunol Rev 74: 129–142

Szpirer J, Defeo-Jones D, Ellis RW, Levan G, Szpirer C (1985) Assignment of three rat cellular RAS oncogenes to chromosomes 1, 4 and X. Somat Cell Mol Genet 11: 93–97

Takatsuki K, Yamaguchi K, Kawano F et al. (1985) Clinical diversity in adult T cell leukemia-lymphoma. Cancer Res 45: 4644 S–4645 S (Suppl)

Tanabe T, Nukada T, Nishikawa Y, Sugimoto K, Suzuki H, Takahashi H, Noda M, Haga T, Ichiyama A, Kangawa K, Minamino N, Mastuo H, Numa S (1985) Primary structure of the alpha-subunit of transducing and its relation to ras proteins. Nature 315: 242–245

Taub R, Kirsch I, Morton C, Lenoir G, Swan D, Tronick S, Aaronson S, Leder P (1982) Translocation of the c-myc gene into the immunoglobulin heavy chain locus in human Burkitt lymphoma and murine plasmacytoma cells. Proc Natl Acad Sci USA 79: 7837–7841

Teich N (1982) Taxonomy of retroviruses. In: Weiss R, Teich N, Varmus H et al. (eds) RNA tumor viruses, vol 1(edn 2). Cold Spring Harbor Laboratory. New York, pp 25–207

Teich N, Wyke J, Mak T et al. (1982) Pathogenesis of retrovirus induced disease. In: Weiss R, Teich N, Varmus H et al. (eds) Molecular biology of tumor viruses. RNA tumor viruses. Cold Spring Harbor, New York, pp 785–998

Testa JR, Parsa NZ, Le Beau MM, Vande Woude, GF (1988) Localization of the proto-oncogene MOS to 8q11-q12 by in situ chromosomal hybridization. Genomics 3: 44–47

Thomas CY, Coffin JM (1982) Genetic alterations of RNA leukemia viruses associated with the development of spontaneous thymic leukemia in AKR/J mice. J Virol 43: 416–426

Thomas CY, Boykin BJ, Famulari NG, Coppola MA (1986) Association of recombinant murine leukemia viruses of the class II genotype with spontaneous lymphomas in CWD mice. J Virol 58: 314–323

Thompson CB, Challoner PB, Neiman PE, Groudine M (1986) Expression of the c-myb proto-oncogene during cellular proliferation. Nature 319: 374–380

Thornell A, Hallberg B, Grundström T (1988) Differential protein binding in lymphocytes to a sequence in the enhancer of the mouse retrovirus SL3-3. Mol Cell Biol 8: 1625–1637

Toda T, Uno I, Ishikawa T, Powers S, Kataoka T, Broek D, Cameron S, Broach J, Mastumoto K, Wigler M (1985) In yeast, RAS proteins are controlling elements of adenylate cyclase. Cell 40: 27–36

Tong L, De Vos AM, Milburn MV, Jancarik J, Naguchi S, Nishimura S, Miura K, Ohtsuka E, Kim SH (1989) Structural differences between a ras oncogene protein and the normal protein. Nature 337: 90–93

Trahey M, McCormick F (1987) A cytoplasmic protein stimulates normal N-ras p21 GTPase, but does not affect oncogenic mutants. Science 238: 542–545

Trahey M, Milley RJ, Cole GE, Innis M, Paterson H, Marshall CJ, Hall A, McCormick F (1987) Biochemical and biological properties of the human N-ras p21 protein. Mol Cell Biol 7: 541–544

Treisman R (1985) Transient accumulation of c-fos RNA following serum stimulation requires a conserved 5' element and c-fos sequences. Cell 42: 889–902

Treisman R (1986) Identification of a protein-binding site that mediated transcriptional response of the c-fos gene to serum factors. Cell 46: 567–574

Tsichlis PN (1987) Oncogenesis by Moloney murine leukemia virus. Anticancer Res 7: 171–180

Tsichlis PN, Coffin JM (1979) Role of the C region in relative growth rates of endogenous and exogenous avian oncoviruses. Cold Spring Harbor Symp Quant Biol 44: 1123–1132

Tsichlis PN, Coffin JM (1980) Recombinants between endogenous and exogenous avian tumor viruses: role of the C region and other portions of the genome in the control of replication and transformaton. J Virol 33: 238–249

Tsichlis PN, Bear SE, Tucker HSG, Schwartz RS (1979) Genetic resistance to ecotropic type-C RNA viruses. In: Immune mechanisms and disease. Academic, New York, pp 101–130

Tsichlis PN, Donehower L, Hager G, Zeller N, Malavarca R, Astrin S, Skalka AM (1982) Sequence comparison of the crossover region of an oncogenic avian retrovirus recombinant and its non-oncogenic parent: genetic regions that control growth rate and oncogenic potential. Mol Cell Biol 2: 1331–1338

Tsichlis PN, Strauss PG, Hu LF (1983) Two common regions for proviral DNA integration in MoMuLV-induced rat thymic lymphomas. Nature 302: 445–449

Tsichlis PN, Strauss PG, Kozak C (1984) A cellular DNA region involved in the induction of thymic lymphomas (Mlvi-2) maps to mouse chromosome 15. Mol Cell Biol 4: 997–1000

Tsichlis PN, Lohse MA, Szpirer C, Szpirer J, Levan G (1985a) Cellular DNA regions involved in the induction of rat thymic lymphomas (Mlvi-1, Mlvi-2, Mlvi-3, and c-myc) represent independent loci as determined by their chromosomal map location in the rat. J Virol 56: 938–942

Tsichlis PN, Strauss PG, Lohse MA (1985b) Concerted DNA rearrangements in Moloney murine leukemia virus-induced thymomas: a potential synergistic relationship in oncogenesis. J Virol 56: 258–267

Tsichlis PN, Shepherd BM, Bear S (1989) Activation of the Mlvi-1/mis-1/pvt-1 locus in Moloney murine leukemia virus induced T cell lymphomas. Proc Natl Acad Sci USA 86: 5487–5491

Tsichlis PN, Lee JS, Bear SE, Lazo PA, Patriotis CP, Gustafson E, Shinton S, Jenkins NA, Copeland NG, Huebner K, Croce C, Levan G, Hanson C (1990) Activation of multiple genes by provirus integration in the Mlvi-4 locus in T cell lymphomas induced by Moloney murine leukemia virus. J Virol 64: 2236–2244

Tsichlis PN, Bear SE (1991) Infection by mink cell focus forming (MCF) viruses confers interleukin-2 (IL-2) independence to an IL-2 dependent rat T cell lymphoma line. Proc Natl Acad Sci USA 88: 4611–4615

Tuchinsky RJ, Stanley ER (1985) The regulation of mononuclear phagocyte entry into S phase by the colony stimulating factor CSF-1. J Cell Physicol 122: 221–228

Tucker HSG, Weens J, Tsichlis PN, Schwartz RS, Khiroya R, Donnelly J (1977) Influence of H-2 complex on susceptibility to infection by murine leukemia virus. J Immunol 118: 1239–1243

Uchiyama T, Hori T, Tsudo M, Wano Y, Umadome H, Tamori S, Yodoi J, Maeda M, Sawami H, Uchino H (1985) Interleukin-2 receptor (Tac antigen) expressed on adult T cell leukemia cells. J Clin Invest 76: 446–453

Van Cong N, Moreau-Gachelin F, Ray D, Gross MS, de Tand MF, Tavitian A, Frézal J (1989a) Assignment of SP11 oncogene to chromosome 11 (somatic cell hybrid analysis), region p 11,22 (in situ hybridization). HGM 10, p 1097 (Abstract) Human Genome Mapping 10

Van Cong N, Fichelson S, Gross MS, Sola B, Bordereaux D, de Tand MF, Guilhot S, Gisselbrecht S, Frezal J. Tambourin P (1989b) The human homologues of Fim-1, Fim-2/c, fms, and Fim-3, three retroviral integration regions involved in mouse myeloblastic leukemias, are respectively located on chromosome 6p23, 5q33, and 3q27. Hum Genet 81: 257–263

Van Lohuizen M, Breuer M, Berns A (1989a) N-myc is frequently activated by proviral insertion in MuLV induced T cell lymphomas. EMBO J 8: 133–136

Van Lohuizen M, Verbeek S, Krimpenfort P, Domen J, Saris C, Radaszkiewicz T, Berns A (1989b) Predisposition to lymphomagenesis in pim-1 transcgenic mice: cooperation with c-myc and N- myc in Murine leukemia virus-induced tumors. Cell 56: 673–682

Varmus H (1988) Regulation of HIV and HTLV gene expression. Genes 2: 1055–1062

Vasmel WLF, Zijlstra M, Radaszkiewicz T, Leupers CJM (1988) Major histocompatibility complex class II-regulated immunity to murine leukemia virus protects against early T-but not late B cell lymphomas. J Virol 62: 3156–3166

Veillette A, Horak ID, Bolen JB (1988a) Post-translational alterations of the tyrosine kinase p56^{ICK} in response to activators of protein kinase C. Onc Res 2: 385–401

Veillette A, Horak ID, Horak EM, Bookman MA, Bolen JB (1988b) Alterations of the lymphocyte-specific protein tyrosine kinase p56^{lck} during T-cell activation. Mol Cell Biol 8: 4353–4361

Veillette A, Bookman MA, Horak EM, Bolen JB (1988c) The CD4 and CD8 T cell surface antigens are associated with the internal membrane tyrosine-protein kinase p56^{lck}. Cell 55: 301–308

Velu TJ, Beguinot L, Vass WC, Willingham MC, Merlino GT, Pastan I, Lowy DR (1987) Epidermal growth factor-dependent transformation by a human EGF receptor proto-oncogene. Science 238: 1408–1410

Vijaya S, Steffen DL, Robinson HL (1986) Acceptor sites for retroviral integrations map near DNase I-hypersensitive sites in chromatin. J Virol 60: 683–692

Vijaya S, Steffen DL, Kozak C, Robinson HL (1987) Dsi-1, a region with frequent proviral insertions in Moloney murine leukemia virus-induced rat thymomas. J Virol 61: 1164–1170

Villemur R, Monczak Y, Rassart E, Kozak C, Jolicoeur P (1987) Identification of a new common provirus integration site in gross passage a murine leukemiavirus-induced mouse thymoma DNA. Mol Cell Biol 7: 512–522

Viskochil D, Buchberg AM, Xu G, Cawthon RM, Stevens J, Wolff RK, Culver M, Carey JC, Copeland NG, Jenkins NA, White R, O'Connell P (1990) Deletions and a translocation interrupt a cloned gene at the neurofibromatosis type 1 locus. Cell 62: 187–192

Vogt PK (1967) Phenotypic mixing in the avian tumor virus group. Virology 32: 708–718

Voronova AF, Sefton BM (1986) Expression of a new tyrosine protein kinase is stimulated by retrovirus promoter insertion. Nature 319. 682–685

Voronova AF, Adler HT, Sefton BM (1987) Two lck transcripts containing different 5' untranslated regions are present in T cells. Mol Cell Biol 7: 4407–4413

Wachsman W, Golde DW, Chen ISY (1986) HTLV and human leukemia: perspective 1986. Semin Hemat 23: 245–256

Waldmann TA, Greene WC, Sarin PS, Satinger C, Blayney DW, Blattner WA, Goldman CK, Bongiovanni K, Sharrow S, Depper JM, Lenord W, Takashi U, Gallo RC (1984) Functional and phenotypic comparison of human T cell leukemia/lymphoma virus positive adult T cell leukemia with human T cell leukemia/lymphoma virus negative Sezary leukemia, and their distinction using anti-Tac J Clin Invest 73: 1711–1717

Wasylyk B, Wasylyk C, Flores P, Begue A, Leprince D, Stehelin D (1980) The c- *ets* proto-oncogenes encode transcription factors that cooperate with c-*fos* and c-*jun* for transcriptional activation. Nature 346: 191–193

Watanabe N, Vande Woude GF, Ikawa Y, Sagata N (1989) Specific protelysis of the c-*mos* protooncogene product by calpain on fertilization of xenopus eggs. Nature 342: 505–511

Watson DK, McWilliams MJ, Papas TS (1988) Molecular organization of the chicken *ets* locus. Virology 164: 99–105

Weiher J, Zonig M, Gruss P (1983) Multiple point mutations affecting the simian virus 40 enhancer. Science 219: 626–631

Weinstein Y, Ihle JN, Lavu S, Reddy EP (1986) Truncation of the c-*myb* gene by a retroviral integration in an interleukin 3-dependent myeloid leukemia cell line. Proc Natl Acad Sci USA 83: 5010–5014

Weinstein Y, Cleveland JL, Askew DS, Rapp UR, Ihle JN(1987) Insertion and truncation of c-*myb* by murine leukemia virus in a myeloid cell line derived from cultures of normal hematopoietic cells. J Virol 61: 2339–2343

Weisbart RH, Golde DW, Clark SC, Wong GG, Gasson JC (1985) Human granulocyte-macrophage colony-stimulating factor is a neutrophil activator. Nature 314: 361–363

Weiss R (1980) Rhabdovirus pseudotypes. In: Bishop DHL (ed) Rhabdoviruses, vol 3. CRC Press, Boca Raton, pp 51–65

Weissman IL, McGrath MS (1982) Retrovirus lymphomagenesis: relationship of normal immune receptors to malignant cell proliferation. Curr Topics Microbiol Immunol 98: 103–112

Wejman JC, Taylor BA, Jenkins NA, Copeland NG (1984) Endogenous xenotropic murine leukemia virus related sequences map to chromosomal regions encoding mouse lymphocyte antigens. J Virol 50: 237–247

Weston K, Bishop JM (1989) Transcriptional activation by the v-*myb* oncogene and its cellular progenitor, c-*myb* . Cell 58: 85–93

Wheeler EF, Askew D, May S, Ihle JN, Sherr CJ (1987) The v-*fms* oncogene induces factor-independent growth and transformation of the interleukin-3-dependent myeloid cell line FDC-P1. Mol Cell Biol 7: 1673–1680

Willingham MC, Pastan I, Shih TY, Scolnick EM (1980) Localization of the *ras* gene product of the Harvey strain of MSV to plasma membrane of transformed cells by electron microscopic immunocytochemistry. Cell 19: 1005–1014

Willingham MC, Banks-Schlegel SP, Pastan IH (1983) Immunocytochemical localization in normal and transformed human cells in tissue culture using a monoclonal antibody to the *ras* protein of the Harvey strain of murine sarcoma virus. Exp Cell Res 149: 141–149

Willumsen BM, Norris K, Papageorge AG, Hubbert NL, Lowy DR (1984) Harvey murine sarcoma virus p21 *ras* protein: biological and biochemical significance of the cysteine nearest the carboxy terminus. EMBO J 3: 2581–2585

Witte ON (1990) Steel locus defines new multipotent growth factor. Cell 63: 5–6

Wolf D, Rotter V (1984) Inactivation of p53 gene expression by an insertion of Moloney murine leukemia virus-like DNA sequences. Mol Cell Biol 4: 1402–1410

Wolf D, Harris N, Rotter V (1984) Reconstitution of p53 expression in a nonproducer Ab-MuLV-transformed cell line by transfection of a functional p53 gene. Cell 38: 119–126

Wong GG, Temple PA, Leary AC, Witek-Giannotti JS, Yang YC, Ciarletta AB, Chung M, Murtha P, Kriz R, Kaufman RJ, Ferenz CR, Sibley BS, Turner KJ, Hewick RM, Clark SC, Yanai N, Yokota H, Yamada M, Saito M, Motoyoshi K, Takaku F (1987) Human CSF-1: molecular cloning and expression of 4-kb cDNA encoding the human urinary protein. Science 235: 1504–1508

Wong GHW, Clark-Lewis I, Harris AW, Schrader JW (1984) Effect of cloned interferon-γ on expression of H-2 and Ia antigens on cell lines of hemopoietic, lymphoid, epithelial, fibroblastic and neuronal origin. Eur J Immunol 14: 52–56

Wong PKY (1990) Moloney murine leukemia virus temperature-sensitive mutants: a model for retrovirus induced neurologic disorders. Curr Topics Microb Immunol 160: 29–60

Wong PKY, Soong MM, MacLeod R, Gallick GE, Yuen PH (1983) A group of temperature-sensitive mutants of Moloney leukemia virus which is defective in cleavage of *env* precursor polypeptide in infected cells also induces hind limb paralysis in newborn CFW/D mice. Virology 125: 513–518

Wong PM, Chung SW, Dunbar CE, Bodine DM, Ruscetti S, Nienhuis AW (1989) Retrovirus-mediated transfer and expression of the interleukin-3 gene in mouse hematopoietic cells result in a myeloproliferative disorder. Mol Cell Biol 9: 798–808

Woolford J, Rothwell V, Rohrschneider L (1985) Characterization of the human c-*fms* gene product and its expression in cells of the monocyte-macrophage lineage. Mol Cell Biol 5: 3458–3466

Woolford J, McAuliffer A, Rohrscheider LR (1988) Activation of the feline c-*fms* proto-oncogene: multiple alterations are required to generate a fully transformed phenotype. Cell 55: 965–977

Ymer S, Tucker WQJ, Sanderson CJ, Hapel AJ, Campbell HD, Young IG (1985) Constitutive synthesis of interleukin-3 by leukemia cell line WEHI-3B is due to retroviral insertion near the gene. Nature 317: 255–258

Yoosook C, Steeves R, Lilly F (1980) *Fv*- 2′-mediated resistance of mouse bone- marrow cells to Friend spleen focus-forming virus infection. Int J Cancer 26: 101–106

Yoshida M, Seiki M (1987) Recent advances in the molecular biology of HTLV-1: transactivation of viral and cellular genes. Ann Rev Immunol 5: 541–559

Yoshida T, Miyagawa K, Odagiri H, Sakamoto H, Little PFR, Terada M, Sugimura T (1987) Genomic sequence of *hst*, a transforming gene encoding a protein homologous to fibroblast growth factors and the *int*-2 encoded protein. Proc Natl Acad Sci USA 84: 7305–7309

Yuan CC, Kan N, Dunn KJ, Papas TS, Blair DG (1989) Properties of a murine retroviral recombinant of avian acute leukemia virus E26: a murine fibroblast assay for v-*ets* function. J Virol 63: 205–215

Zijlstra M, Melief CJM (1986) Virology, genetics and immunology of murine lymphomagenesis. Biochim Biophys Acta 865: 197–231

Subject Index

A-CBF 104
Abelson MuLV 118, 124, 137
Adenovirus E1A 13S protein 108, 137
adenylate cyclase 130, 131
AKR thymoma virus #8 (AKT-8) 144
Akv 102, 104
AKXD-23 mice 117
ALV (Avian leukosis virus) 96, 98, 116, 125
ATL (adult T cell leukemia) 146, 148
avian leukosis virus (ALV)
– B-cell lymphoma 4, 5, 14
– c-*erb*B activation 4, 5, 11, 13, 16
– c-*fos* activation 4, 5
– c-Ha-*ras* 4
– c-*myc* activation 4, 5, 11, 12
– c-*ras* activation 4, 5
– erythroblastosis 4, 5, 16
– *gag* enhancer 14
– LTR promoter occlusion 14
– nephroblastoma 4, 5
– proviral deletion 14
– RAV-1,2 4, 5

B cell lymphomas 120, 121
B-tropic viruses 110
Balb sarcoma virus 144
Balb/c ecotropic virus 120, 121
basophils 126
β-globulin 134
BFU-E 114
BLV (Bovine leukemia virus) 144
breast cancer, human 58
BXH-2 129

c-*erb*B
– erythroblastosis 11, 13, 16
– insertional activation 11, 13, 16
– truncation 16
– transduction 16

c-fos 142, 145, 147
c-jun 142
c-kit 115, 128
c-*myc* 83
– B-cell lymphomas 11, 12
– insertional activation by 11–14
– – avian leukosis virus 11, 12
– – chicken syncytial virus 11, 14
– – feline leukemia virus 11, 12
– – murine leukemia virus 11, 12
– insertional map 11
– structure 11
– T-cell lymphomas 11, 12
calpain 132
cAMP
– control element 147
– responsive gene 147
capsid protein 109
carboxymethylation 131
Cas-Br-E 197
Cas-Br-M-MuLV 118
CD3 132
CD4 132, 133, 145, 148
CD8 123, 132, 133
cdc 2 kinase 132
chicken(s)
– 15₁ chickens 144
– (K28 x 15₁) x K28 chickens 144
– syncytial virus (CSV)
– – B-cell lymphoma 5, 12, 14
– – c-*myc* activation 4, 5, 11, 14, 15
– – p53 inactivation by proviral insertion 17
– – proviral deletion 14
– – reticuloendotheliosis virus (REV) 12
CMV IE2 protein 108
cooperation 57
CTG codon 133
CUG initiation codon 55

CWD mice   100
cyclin   132
cytotoxic T cell (CTL)   111, 113
Czech-2   57

DA3   123
DA34   123
dermis   145
destabilization of mRNA   122
determinants of viral pathogenicity (see
   also viral pathogenicity)   98
diacylglycerol (DAG)   132
disease specificity   98, 100, 102–104
DNA, unintegrated proviral   109
downstream element-1 (DE-1)   147
Drosophila   49
DU5H/LP-BM5   106, 107

E26 retrovirus   135
ecotropic viruses   99, 105, 107
embryogenensis   50
embryonal carcinoma cell lines   50
embryonic stem cells   50
endogenous proviruses   99
endothelial cells, neuronal   107
enhancers   45
envelope glycoprotein   105, 107, 110, 111
eosinophilia   146
eosinophils   126, 127
Eph   129
epidermal growth factor receptor   11, 13,
   16, 128
erythrocytes   126
erythroid cell line 416B   117
erythroleukemia   102–104, 107, 114, 115,
   124, 135, 144
estrogen receptor   133
Ets-1 Exon 7   135
extracellular matrix   49
extramedullary hematopoiesis   106

F-MuLV (Friend murine leukemia
   virus)   103, 106, 110, 119, 124, 129,
   136
FBJ sarcoma virus   144
FBR sarcoma virus   144
feline leukemia virus (FeLV)   65, 96, 144
– endogenous   68
– envelope glycoproteins   69
– FAIDS   74
– receptors   69
– recombination   70
– subgroup   69
– transduction   79

fibroblast growth factor (FGF)   47
Flvi-1   78
FRG mice   110
Friend virus complex   111, 112, 114,
   115

GAL4   134
Gazdar sarcoma virus   132, 144
genetic variants of retroviruses   98, 100,
   104, 106
germ cells, postmitotic   132
glial cells   145
glycosylation   47
gp70   105
granulocytosis   146
growth
– factor, platelet derived (PDGF)   128
– factors, hematopoietic   105, 126, 127
– receptor, epidermal growth factor   128
GTPase   130
– activating protein (GAP)   130

H-2 complex   111, 112
– K end   111
– D end   112
– I region   111, 112
Harvey sarcoma virus   144
HBV X protein   108
hematopoiesis, extramedullary   105
hematopoietic
– growth factors   105, 126, 127
– neoplasms   107, 108, 117
heparin   56
high leukemia mouse strains   98–100,
   106
histamine release   126
HIV
– EN86A   147
– rev gene   108
homologous recombination between integrated proviruses   125
host genes determining viral pathogenicity
   (see also viral
pathogenicity)   108–115
hst   56
HSV-1 ICPO protein   108
HTLV-1 (human T cell lymphotropic
   virus-I)   96, 142, 144–148
– p21$_{rex}$ 145
– repressor X (rex)   145
– transactivator X (tax)   145
human breast cancer   58
hypercalcemia   146
hyperplastic glands   52

immunodeficiency   74
immunosuppressive   105, 106
inbred mice   99
− AKR   97, 99, 100, 111
− C58/J   99
− DBA/2   99, 108, 110
− recombinant inbred mice AKXD   99, 111
incubation period   148
inositol triphosphate (IP3)   131
insertional see also provirus insertion and
   provirus integration
− activation of proto-oncogenes
− − common insertion sites   5, 16
− − CSV strain of reticuloendotheliosis
   virus (REV)   12
− − mechanisms
− − − enhancer insertion   9, 10, 12
− − − influence of host factors   10–13
− − − influence of viral elements   13, 14
− − − 3' LTR promotion   8, 9, 12, 13
− − − 5' LTR promotion   8, 9, 13
− − − promoter insertion   8, 9, 12, 13
− − − readthrough activation   8, 9, 13
− frequency   3
− inactivation
− − collagen type I gene   3
− − dilute coat color   2
− − hairless   3
− − host genes   2
− − hprt   3
− − p53   17
− − slow feathering   3
− − v-src   3
− mutagenesis   27–30, 32, 33, 38, 39, 78,
   98, 105, 115, 116, 123, 124, 127, 144
− − dilute   31, 32, 38
− − hairless   31, 32, 38
− − hprt   30
− − recessive lethal mutations   28–31,
   37, 38
− − − Mov-13   28, 38
− − − Mov-34   28, 29, 38
− − − Srev-5   37, 38
− specificity/randomnes
int
− 1   45
− 2   45
− 3   47
− amplification   58
− genes, cooperation   57
intracisternal A particles (IAP)   127
isoprenylation   131

Kirsten sarcoma virus   144

L3T4   105
latency period   100, 103, 131
Lentivirinae   96
LTR
− enhancer   9, 10, 12, 76
− promoter
− − adhesion   12, 14, 15
− − competition   15
− − epigenic suppression   15
− − occlusion 12
lymphomas   97, 99, 103, 112, 113,
   115–117, 119, 122–124, 131, 138, 141,
   142, 144, 148
− B cell   131, 136
− nonthymic   99
− pre B   124, 126, 131, 137
− T cell   97, 99, 103, 112, 113, 115, 117,
   119, 122, 123, 125, 131, 138, 141, 142
− thymic   125

mac   133
macrophages   126–129, 136
MAIDS virus   107
mammary
− adenocarcinomas   123, 142
− gland hyperplasia   56
− tumors   43–61
Marek's disease virus   108
mast cell   126
maturation promoting factor (MPF)   132
MCF
− 1233   112
− viruses (see also polytropic viruses)
   99, 100, 104, 105, 110, 113
− − class I   99, 100, 111, 132
− − class II   99, 100, 111, 132
megakaryocytes   126
mesoderm induction   55
MLFV   120, 121
MMTV   44, 123, 143
− genome, enhancers   45
MoBA-I   106
Moloney
− MSV (MoMSV)   100, 104, 135
− MuLV (MoMuLV)   97, 102, 103, 111,
   114, 117–119, 121, 123, 125, 138, 141,
   142, 148
− − mutants   106, 107
− sarcoma virus   144
mouse; mice
− AKXD-23   117
− CBA/N   114
− chimera   28, 30, 38
− CWD   100

mouse
- embryo   28–30, 32–39
- embryonic stem cells   28–30, 32, 38, 39
- FRG   110
- germline   28–39
- imbred   99
- - AKR   97, 99, 100, 111
- - C58/J   99
- - DBA/2   99, 108, 110
- - recombinant inbred mice AKXD   99, 111
- ovarian transplantation   34–36
- RF/J   33–37, 39
- SWR/J   33–37, 39
mRNA, destabilization   122
Mus musculus   110, 141
Mus spretus   142
myeloblastic leukemia   120, 121
myeloid
- cell lines   118, 123
- - DA-1   120, 121
- - DA-3   120, 121, 123
- - DA-34   120, 121, 123
- - NFS 58   118, 120, 121
- - NFS 60   120, 121
- - NFS 78   118, 120, 121
- neoplasms   123, 136
- precursor cell lines   120, 121
- - DIND-1   120, 121
- - DIND-4   120, 121
- - DIND-5   120, 121
- - DIND-9   120, 121
myelomonocytic neoplasm   128

N tropic viruses   110
NB tropic viruses   110
neoplasm(s)
- myelomonocytic   128
- hematopoietic   11, 107, 108, 117
- myeloid   123, 136
nervous system   50
neurofibromatosis   129
- like syndrome   145
neuronal endothelial cells   107
neurotoxicity   107
neurotropic viruses   107
neutrophils   126, 127
NF-1 binding site   103
NF-kB   146, 147
NFSXDBA/2   111
nonacute retroviruses   96, 115, 147
nonthymic lymphomas   99
nontransforming retroviruses   96, 115, 144

nuclear
- factor-1 (NF-1)   103
- uptake   55

oncogene(s)   79, 142
- cooperating   142
- proto-oncogenes   3
- recessive   123
- transduction   16, 79, 143
oncogenesis   96–98, 100, 104–106, 108, 114–116, 124, 133, 138, 140–144, 149
oncovirinae   96
oocytes   102, 132, 136
orientation of proviruses   45

p15E   105, 129
palmitoylation   131
PEA3 site   135
PHA   147
phenotype
- mixing   97
- T cell lymphomas   97
pheochromocytoma cells (PC12)   131
phospholipase C (PLC)   131
plasmacytoid lymphosarcomas   118
plasmacytoma cell lines   120, 121
- NSI   120, 121
- MOPC21   120, 121
platelet derived growth factor (PDGF)   128
plus strand RNA viruses   96
PMA   147
polytropic viruses   100
- Pmv-25   99
- Pmv-40   110, 111
postmitotic germ cells   132
Pr-RSV-B   98
pre-B cell line   124
pre-T cells   126
preleukemic phase   98–100, 106
protease   105, 132
protein kinase C   131, 144
provirus insertion
- affected genes   126–141
- - growth factors   126–128
- - - colony stimulation factor-1   128
- - - granulocyte-macrophage colony stimulation factor (GM-CSF)   127
- - - interleukin 3 (IL-3)   126
- - growth factor receptors   11, 13, 16, 128–130
- - - c-fms/fim-2   128
- - - Evi-2   129, 130
- - loci marking genes, un- defined   138–141

– – – *ahi*-1   139, 140
– – – *dsi*-1   139, 140
– – – *fim*-1   139, 140
– – – *fis*-1   139, 140
– – – *fli*-1   139
– – – *gin*-1   139, 140
– – – *Mlvi*-2   139, 140
– – – *Mlvi*-3   139, 140
– – – *pim*-2   139, 140
– – nuclear proteins/transcription factors   133–137
– – – *Evi*-1   117, 118, 122, 123, 136
– – – *Hox*-2.4   117, 136, 137
– – – *myb*   118, 134
– – – *myc*   133, 136, 143
– – – *p53*   124, 125, 130, 137
– – – *Sfpi*-1/*Spi*-1/*PU*-1   136
– – – *Tpl*-1/*Ets*-1 (see also *Ets*-1 binding site)   134
– – partially characterized genes   137–138
– – – *Gfi*-1   138, 141
– – – *Mlvi*-1   122, 123, 137, 138, 143
– – – *Mlvi*-4   117, 119, 122, 123, 137, 138, 143
– – – *Tpl*-2   138, 141
– – signal transduction   130–133
– – – *c-Ha-ras*   117, 125, 130
– – – *c-Ki-ras*   117, 130
– – – *Lck*   117, 132
– – – *mos*   132, 144
– – – *pim*-1   122, 131, 143
– effects   117–125
– – activation of neighboring genes   123
– – dominant negative mutations   123, 124
– – enhancer insertion   119, 122, 127
– – gene inactivation   123
– – long distance gene activation   122
– – oncogenes, recessive   123
– – promoter insertion   117, 119, 124, 138
– – RNA message, stability   122
– – synthesis of abnormal gene product   119
– – truncation of gene   119
provirus integration (see also insertional mutagenesis)   98, 109, 115–117, 119, 122–124, 126, 129, 130, 136, 138, 141–143, 149
– spontanous recurrent   141
proviruses
– endogenous   99
– homologous recombination between integrated proviruses   125

– orientation   45
pseudotyping   107

radiation leukemia virus (RadLV)   112
Rasheed sarcoma virus   144
Rauscher MuLV   120, 121
RAV-1   144
RAV-O   98
Rb gene   137
receptor(s)   49
– estrogen   133
– FeLV   69
– growth
– – epidermal   128
– – provirus insertion   128–130
– T cell 105, 132, 147
recessive oncogenes   123
Ret   128
retroviruses (see also viruses)
– E26   135
– genetic variants   98, 100, 104, 106
– nonacute   96, 115, 147
– nontransforming   96, 115, 144
– pathogenesis 96, 98, 141, 149
– transposable elements   27, 28, 30, 32, 39
– type B   96
– type C   96–98, 100, 102, 107
rhabdovirus   107
RNA
– enhancer   15
– mRNA, destabilization   122
– viruses, plus strand   96
Rous-associated virus-1 (RAV-1)   4

S/A-CBF   104
sarcoma   100, 132, 143, 144
– virus
– – Balb   144
– – FBJ   144
– – FBR   144
– – Gazdar   132, 144
– – Harvey   144
– – Kirsten   144
– – Moloney   144
– – Rasheed   144
Schwann cells   145
20γ SDH   114
Sea   128
secretory proteins.   47
segment polarity genes   51
serine/threonine kinase   131, 132
SFFV   105–107, 136
signal peptide   53

Sjörgen's like syndrome   145
SL3-3 virus   100, 102
Sp-1   147
spermatids   50, 132
spongiform encephalophathy   107
Spumavirinae   96
stages of viral leukemogenesis   97
suramin   49
SV40   108, 137

T cell
– activation   132, 135, 147
– dependent antigens   113
– development   97
– lines   104
– – lymphoma   120, 121, 138, 141
– – – 6889   120, 121
– – – 6890   120, 121
– – – BW5147   122
– – – DA-2   120, 121
– – – LSTRA   120, 121
– – – SL12.4.10   123
– – – Thy-19   120, 121
– lymphomas   125
– – phenotype   97
– receptor   105, 132, 147
testis   50
thymic lymphomas   125
thymomagenic viruses   100
thymotropic viruses   98–100
TNF-γ   147
transactivation   107, 108, 144–149
– c-fos   142, 145, 147
– GM-CSF   126, 127, 145, 146
– IL-2   138, 141, 145–147
– IL-2Rγ   145–147
transduction   16
transmembrane protein   105, 129
transposable elements   27, 28, 39
– retrovirus   27, 28, 30, 32, 39
– – ecotropic   30–37
– – – Emv loci   31, 33–37, 39
– – – MoMuLV   28, 29
– – polytropic   32
transposon tagging   47
tropical spastic paraparesis (TSP)   96
tumor(s)
– induction   97, 98, 108, 109, 112, 116,
  144, 148
– mammary   43–61
– progression   97, 135, 138, 141
type B retroviruses   96
type C retroviruses   96–98, 100, 102, 107
tyrosine kinase   115, 128, 130, 132, 144

unintegrated proviral DNA   109

v-myc   85
vesicular stomatitis virus (VSV)   107
viral
– enhancer   102
– – LTR   98–100, 102–105, 107, 110,
  115, 117, 118, 122, 125, 135,
  144–149
– – polyoma   102, 135, 137
– – SV40   102
– interactions   107
– leukemogenesis, stages   97
– pathogenicity, determined by host
  genes   108–115
– – immune response   111
– – – Fv-3   114
– – – hr   113
– – – Rfv-1   113
– – – Rfv-2   113
– – – Rfv-3   113
– – – Rgv-1   111, 112
– – – X-linked immunodeficiency (CBA/N
  mice)   114
– – target pool, size   114
– – – F   115
– – – Nu   115
– – – Fv-2   114
– – – Sl   115
– – – W   115
– – type of disease   115
– – – Fv-5   115
– – – Fv-6   115
– – virus replication   108
– – – Fv-1   99, 109
– – – Fv-4/Akrv-1   110
– – – Fv-6/Rmcf   110, 111
– – – Gv-1   111
– – – Gv-2   111
– – – Rgv-2   111
– – – Srv-2   111
– pathogenicity, determined by viral
  genes   98–107
– – env   98, 99, 104, 105, 107, 110,
  145
– – LTR   98–100, 102–105, 107, 110,
  115, 117, 118, 122, 125, 135,
  144–149
– – – CArG box   103
– – – CAT box   102
– – – conserved sequences   102, 135,
  138
– – – direct repeats   102, 103

– – – enhancer core   100, 102–104
– – – Ets-1 binding site (see also Tpl-
  1/Ets-1)   102, 104 $SCHREIB
– – – GC rich sequence   102
– – – GRE   103
– – – inverted repeats   102
– – – Lva site   103
– – – Lvb/Lvt site   102, 104
– – – Lvc site   103
– – – TATA box   102, 104
– – – upstream control region
  (UCR)   102, 104
viruses
– B-tropic   110
– ecotropic   99, 105, 107
– MCF   99, 100, 104, 105, 110, 113
– – class I 99, 100, 111, 132
– – class II   99, 100, 111, 132
– N tropic   110
– NB tropic viruses   110

– neurotropic   107
– plus strand RNA   96
– polytropic   100
– – Pmv-25   99
– – Pmv-40   110, 111
– thymomagenic   100
– thymotropic   98–100
– xenotropic   99   --Bxv-1   99
Von Recklinghansen   129

WEH1-3B   127, 136
wingless   49
Wnt-1   45
Wnt-2   52
Wnt-3   52

X-chromosome   39
Xenopus embryos   50
xenotropic viruses   99
– Bxv-1   99

Current Topics in Microbiology and Immunology

Volumes published since 1986 (and still available)

Vol. 126: **Fleischer, Bernhard; Reimann, Jörg; Wagner, Hermann (Ed.):** Specificity and Function of Clonally Developing T-Cells. 1986. 60 figs. XV, 316 pp.
ISBN 3-540-16501-0

Vol. 127: **Potter, Michael; Nadeau, Joseph H.; Cancro, Michael P. (Ed.):** The Wild Mouse in Immunology. 1986. 119 figs. XVI, 395 pp. ISBN 3-540-16657-2

Vol. 128: 1986. 12 figs. VII, 122 pp.
ISBN 3-540-16621-1

Vol. 129: 1986. 43 figs., VII, 215 pp.
ISBN 3-540-16834-6

Vol. 130: **Koprowski, Hilary; Melchers, Fritz (Ed.):** Peptides as Immunogens. 1986. 21 figs. X, 86 pp. ISBN 3-540-16892-3

Vol. 131: **Doerfler, Walter; Böhm, Petra (Ed.):** The Molecular Biology of Baculoviruses. 1986. 44 figs. VIII, 169 pp.
ISBN 3-540-17073-1

Vol. 132: **Melchers, Fritz; Potter, Michael (Ed.):** Mechanisms in B-Cell Neoplasia. Workshop at the National Cancer, Institute, National Institutes of Health, Bethesda, MD, USA, March 24–26, 1986. 1986. 156 figs. XII, 374 pp. ISBN 3-540-17048-0

Vol. 133: **Oldstone, Michael B. (Ed.):** Arenaviruses. Genes, Proteins, and Expression. 1987. 39 figs. VII, 116 pp.
ISBN 3-540-17246-7

Vol. 134: **Oldstone, Michael B. (Ed.):** Arenaviruses. Biology and Imminotherapy. 1987. 33 figs. VII, 242 pp.
ISBN 3-540-14322-6

Vol. 135: **Paige, Christopher J.; Gisler, Roland H. (Ed.):** Differentiation of B Lymphocytes. 1987. 25 figs. IX, 150 pp.
ISBN 3-540-17470-2

Vol. 136: **Hobom, Gerd; Rott, Rudolf (Ed.):** The Molecular Biology of Bacterial Virus Systems. 1988. 20 figs. VII, 90 pp.
ISBN 3-540-18513-5

Vol. 137: **Mock, Beverly; Potter, Michael (Ed.):** Genetics of Immunological Diseases. 1988. 88 figs. XI, 335 pp.
ISBN 3-540-19253-0

Vol. 138: **Goebel, Werner (Ed.):** Intracellular Bacteria. 1988. 18 figs. IX, 179 pp.
ISBN 3-540-50001-4

Vol. 139: **Clarke, Adrienne E.; Wilson, Ian A. (Ed.):** Carbohydrate-Protein Interaction. 1988. 35 figs. IX, 152 pp.
ISBN 3-540-19378-2

Vol. 140: **Podack, Eckhard R. (Ed.):** Cytotoxic Effector Mechanisms. 1989. 24 figs. VIII, 126 pp. ISBN 3-540-50057-X

Vol. 141: **Potter, Michael; Melchers, Fritz (Ed.):** Mechanisms in B-Cell Neoplasia 1988. Workshop at the National Cancer Institute, National Institutes of Health, Bethesda, MD, USA, March 23–25, 1988. 1988. 122 figs. XIV, 340 pp.
ISBN 3-540-50212-2

Vol. 142: **Schüpach, Jörg:** Human Retrovirology. Facts and Concepts. 1989. 24 figs. 115 pp. ISBN 3-540-50455-9

Vol. 143: **Haase, Ashley T.; Oldstone Michael B. A. (Ed.):** In Situ Hybridization 1989. 22 figs. XII, 90 pp.
ISBN 3-540-50761-2

Vol. 144: **Knippers, Rolf; Levine, A. J. (Ed.):** Transforming. Proteins of DNA Tumor Viruses. 1989. 85 figs. XIV, 300 pp.
ISBN 3-540-50909-7

Vol. 145: **Oldstone, Michael B. A. (Ed.):** Molecular Mimicry. Cross-Reactivity between Microbes and Host Proteins as a Cause of Autoimmunity. 1989. 28 figs. VII, 141 pp. ISBN 3-540-50929-1

Vol. 146: **Mestecky, Jiri; McGhee, Jerry (Ed.):** New Strategies for Oral Immunization. International Symposium at the University of Alabama at Birmingham and Molecular Engineering Associates, Inc. Birmingham, AL, USA, March 21–22, 1988. 1989. 22 figs. IX, 237 pp. ISBN 3-540-50841-4

Vol. 147: **Vogt, Peter K. (Ed.):** Oncogenes. Selected Reviews. 1989. 8 figs. VII, 172 pp. ISBN 3-540-51050-8

Vol. 148: **Vogt, Peter K. (Ed.):** Oncogenes and Retroviruses. Selected Reviews. 1989. XII, 134 pp. ISBN 3-540-51051-6

Vol. 149: **Shen-Ong, Grace L. C.; Potter, Michael; Copeland, Neal G. (Ed.):** Mechanisms in Myeloid Tumorigenesis. Workshop at the National Cancer Institute, National Institutes of Health, Bethesda, MD, USA, March 22, 1988. 1989. 42 figs. X, 172 pp. ISBN 3-540-50968-2

Vol. 150: **Jann, Klaus; Jann, Barbara (Ed.):** Bacterial Capsules. 1989. 33 figs. XII, 176 pp. ISBN 3-540-51049-4

Vol. 151: **Jann, Klaus; Jann, Barbara (Ed.):** Bacterial Adhesins. 1990. 23 figs. XII, 192 pp. ISBN 3-540-51052-4

Vol. 152: **Bosma, Melvin J.; Phillips, Robert A.; Schuler, Walter (Ed.):** The Scid Mouse. Characterization and Potential Uses. EMBO Workshop held at the Basel Institute for Immunology, Basel, Switzerland, February 20–22, 1989. 1989. 72 figs. XII, 263 pp. ISBN 3-540-51512-7

Vol. 153: **Lambris, John D. (Ed.):** The Third Component of Complement. Chemistry and Biology. 1989. 38 figs. X, 251 pp. ISBN 3-540-51513-5

Vol. 154: **McDougall, James K. (Ed.):** Cytomegaloviruses. 1990. 58 figs. IX, 286 pp. ISBN 3-540-51514-3

Vol. 155: **Kaufmann, Stefan H. E. (Ed.):** T-Cell Paradigms in Parasitic and Bacterial Infections. 1990. 24 figs. IX, 162 pp. ISBN 3-540-51515-1

Vol. 156: **Dyrberg, Thomas (Ed.):** The Role of Viruses and the Immune System in Diabetes Mellitus. 1990. 15 figs. XI, 142 pp. ISBN 3-540-51918-1

Vol. 157: **Swanstrom, Ronald; Vogt, Peter K. (Ed.):** Retroviruses. Strategies of Replication. 1990. 40 figs. XII, 260 pp. ISBN 3-540-51895-9

Vol. 158: **Muzyczka, Nicholas (Ed.):** Viral Expression Vectors. 1992. Approx. 20 figs. Approx. XII, 190 pp. ISBN 3-540-52431-2

Vol. 159: **Gray, David; Sprent, Jonathan (Ed.):** Immunological Memory. 1990. 38 figs. XII, 156 pp. ISBN 3-540-51921-1

Vol. 160: **Oldstone, Michael B. A.; Koprowski, Hilary (Ed.):** Retrovirus Infections of the Nervous System. 1990. 16 figs. XII, 176 pp. ISBN 3-540-51939-4

Vol. 161: **Racaniello, Vincent R. (Ed.):** Picornaviruses. 1990. 12 figs. X, 194 pp. ISBN 3-540-52429-0

Vol. 162: **Roy, Polly; Gorman, Barry M. (Ed.):** Bluetongue Viruses. 1990. 37 figs. X, 200 pp. ISBN 3-540-51922-X

Vol. 163: **Turner, Peter C.; Moyer, Richard W. (Ed.):** Poxviruses. 1990. 23 figs. X, 210 pp. ISBN 3-540-52430-4

Vol. 164: **Bækkeskov, Steinnun; Hansen, Bruno (Ed.):** Human Diabetes. 1990. 9 figs. X, 198 pp. ISBN 3-540-52652-8

Vol. 165: **Bothwell, Mark (Ed.):** Neuronal Growth Factors. 1991. 14 figs. IX, 173 pp. ISBN 3-540-52654-4

Vol. 166: **Potter, Michael; Melchers, Fritz (Ed.):** Mechanisms in B-Cell Neoplasia 1990. 143 figs. XIX, 380 pp. ISBN 3-540-52886-5

Vol. 167: **Kaufmann, Stefan H. E. (Ed.):** Heat Shock Proteins and Immune Response. 1991. 18 figs. IX, 214 pp. ISBN 3-540-52857-1

Vol. 168: **Mason, William S.; Seeger, Christoph (Ed.):** Hepadnaviruses. Molecular Biology and Pathogenesis. 1991. 21 figs. X, 206 pp. ISBN 3-540-53060-6

Vol. 169: **Kolakofsky, Daniel (Ed.):** Bunyaviridae. 1991. 34 figs. X, 256 pp. ISBN 3-540-53061-4

Vol. 170: **Compans, Richard W. (Ed.):** Protein Traffic in Eukaryotic Cells. Selected Reviews. 1991. 14 figs. X, 186 pp. ISBN 3-540-53631-0